Communications in Computer and Information Science 1865

Editorial Board Members

Rationale

The CCIS series is devoted to the publication of proceedings of computer science conferences. Its aim is to efficiently disseminate original research results in informatics in printed and electronic form. While the focus is on publication of peer-reviewed full papers presenting mature work, inclusion of reviewed short papers reporting on work in progress is welcome, too. Besides globally relevant meetings with internationally representative program committees guaranteeing a strict peer-reviewing and paper selection process, conferences run by societies or of high regional or national relevance are also considered for publication.

Topics

The topical scope of CCIS spans the entire spectrum of informatics ranging from foundational topics in the theory of computing to information and communications science and technology and a broad variety of interdisciplinary application fields.

Information for Volume Editors and Authors

Publication in CCIS is free of charge. No royalties are paid, however, we offer registered conference participants temporary free access to the online version of the conference proceedings on SpringerLink (http://link.springer.com) by means of an http referrer from the conference website and/or a number of complimentary printed copies, as specified in the official acceptance email of the event.

CCIS proceedings can be published in time for distribution at conferences or as postproceedings, and delivered in the form of printed books and/or electronically as USBs and/or e-content licenses for accessing proceedings at SpringerLink. Furthermore, CCIS proceedings are included in the CCIS electronic book series hosted in the SpringerLink digital library at http://link.springer.com/bookseries/7899. Conferences publishing in CCIS are allowed to use Online Conference Service (OCS) for managing the whole proceedings lifecycle (from submission and reviewing to preparing for publication) free of charge.

Publication process

The language of publication is exclusively English. Authors publishing in CCIS have to sign the Springer CCIS copyright transfer form, however, they are free to use their material published in CCIS for substantially changed, more elaborate subsequent publications elsewhere. For the preparation of the camera-ready papers/files, authors have to strictly adhere to the Springer CCIS Authors' Instructions and are strongly encouraged to use the CCIS LaTeX style files or templates.

Abstracting/Indexing

CCIS is abstracted/indexed in DBLP, Google Scholar, EI-Compendex, Mathematical Reviews, SCImago, Scopus. CCIS volumes are also submitted for the inclusion in ISI Proceedings.

How to start

To start the evaluation of your proposal for inclusion in the CCIS series, please send an e-mail to ccis@springer.com.

Alvaro David Orjuela-Cañón · Jesus A Lopez ·
Julián David Arias-Londoño

Editors

Applications of Computational Intelligence

6th IEEE Colombian Conference, ColCACI 2023
Bogota, Colombia, July 26–28, 2023
Revised Selected Papers

 Springer

Editors
Alvaro David Orjuela-Cañón (iD)
School of Medicine and Health Sciences
Universidad del Rosario
Bogotá, Colombia

Jesus A Lopez (iD)
Automatics and Electronics Department
Universidad Autónoma de Occidente
Cali, Colombia

Julián David Arias-Londoño (iD)
Universidad Politécnica de Madrid
Madrid, Spain

ISSN 1865-0929   ISSN 1865-0937 (electronic)
Communications in Computer and Information Science
ISBN 978-3-031-48414-8   ISBN 978-3-031-48415-5 (eBook)
https://doi.org/10.1007/978-3-031-48415-5

This Springer imprint is published by the registered company Springer Nature Switzerland AG
The registered company address is: Gewerbestrasse 11, 6330 Cham, Switzerland

Paper in this product is recyclable.

Preface

Computational intelligence (CI) solutions are increasingly employed to engineering problems in the Latin America (LA) region. LA scientists have focused their efforts on CI approaches as a way to deal with problems of interest for the international community but also of great impact in the LA region. Many different problems, including optimization of energy and transportation systems, computer-aided medical diagnoses, bioinformatics, mining of massive data sets, robotics and automatic surveillance systems, among many others, are commonly addressed in this part of the world, and these applications have great potential in developing countries.

The Colombian chapter of the IEEE Computational Intelligence Society (IEEE CIS) and the IEEE Colombia section chose Bogota, Colombia as an in-person venue for the IEEE Colombian Conference on Applications of Computational Intelligence (IEEE Col-CACI 2023). As with previous editions, the conference sought to be the most important event in computational intelligence and related fields in Colombia, bringing together academia, science and industry.

This sixth edition, IEEE ColCACI 2023, was fortified with contributions from scientists, engineers and practitioners working on both applications and theory of CI techniques. We received 42 papers by authors from three Andean countries (Colombia, Ecuador and Peru) and countries from other parts of the globe, such as Brazil and India. The conference had 23 oral presentations applying in-person mode and also other ways to share ideas and talk about CI topics. Thus the event was an international forum for CI researchers and practitioners to share their most-recent advancements and results. This post-proceedings, with revised selected papers, includes the best 11 papers presented as extended versions of work exhibited at the conference. And we are committed to continue working to offer excellent IEEE ColCACI events in future.

Finally, we would like to thank the IEEE Colombia section, the IEEE Computational Intelligence Society Colombian chapter, the IEEE Computational Intelligence Society, the Universidad Autónoma de Occidente, the Universidad del Rosario and its Claustro Campus, the Universidad de Antioquia, and Springer for their support. Also, special thanks to all the volunteers, participants, and the whole crew who worked together to create a successful conference.

September 2023

Alvaro David Orjuela-Cañón
Julián David Arias-Londoño
Jesus A. Lopez

Organization

Organizers

General Chairs

Jesús Alfonso López Sotelo	Universidad Autónoma de Occidente, Colombia
Alvaro David Orjuela-Cañón	Universidad del Rosario, Colombia

Program Committee Chairs

Julián David Arias-Londoño	Universidad de Antioquia, Colombia
Juan Carlos Figueroa-García	Universidad Distrital Francisco José de Caldas, Colombia

Publication Chair

Alvaro David Orjuela-Cañón	Universidad del Rosario, Colombia

Financial Chair

Leidy Ferro	Ecopetrol, Colombia

Webmaster

Fabian Martinez	IEEE Colombia, Colombia

Program Committee

Alvaro David Orjuela Cañón	Universidad del Rosario, Colombia
Jesus A. Lopez	Universidad Autónoma de Occidente, Colombia
Julián David Arias Londoño	Universidad de Antioquia, Colombia
Juan Carlos Figueroa García	Universidad Distrital Francisco José de Caldas, Colombia
Danton Ferreira	Universidade Federal de Lavras, Brazil
Efren Gorrostieta	Universidad Autónoma de Queretaro, Mexico
Cristian Rodríguez Rivero	UC Davis, USA
Jose Alfredo Costa	Universidade Federal do Rio Grande do Norte, Brazil
Javier Mauricio Antelis	Instituto Tecnológico de Monterrey, Mexico

Leonardo Forero Mendoza	Universidade Estadual do Rio de Janeiro, Brazil
Carmelo Bastos Filho	Universidade de Pernambuco, Brazil
Edgar Sánchez	CINVESTAV, Unidad Guadalajara, Mexico
Guilherme Alencar Barreto	Universidade Federal do Ceará, Brazil
Gonzalo Acuña Leiva	Universidad de Santiago de Chile, Chile
Carlos Alberto Cobos Lozada	Universidad del Cauca, Colombia
Juan Bernardo Gómez Mendoza	Universidad Nacional de Colombia, Sede Manizales, Colombia
Diego Peluffo Ordóñez	Universidad Técnica del Norte, Ecuador
Gerardo Muñoz Quiñones	Universidad Distrital Francisco José de Caldas, Colombia
Alvaro David Orjuela Cañón	Universidad del Rosario, Colombia
Jorge Eliécer Camargo Mendoza	Universidad Antonio Nariño, Colombia
Claudia Victoria Isaza Narvaez	Universidad de Antioquia, Colombia
Sandra Esperanza Nope Rodríguez	Universidad del Valle, Colombia
Jesús Alfonso López Sotelo	Universidad Autónoma de Occidente, Colombia
Cesar Hernando Valencia Niño	Universidad Santo Tomás, Sede Bucaramanga, Colombia
Miguel Melgarejo Rey	Universidad Distrital Francisco José de Caldas, Colombia
Wilfredo Alfonso Morales	Universidad del Valle, Colombia
Diana Consuelo Rodríguez	Universidad del Rosario, Colombia
Humberto Loaiza	Universidad del Valle, Colombia
Eduardo Francisco Caicedo Bravo	Universidad del Valle, Colombia
Alexander Molina Cabrera	Universidad Tecnológica de Pereira, Colombia
Luiz Pereira Caloba	Universidade Federal de Rio de Janeiro, Brazil
Leonardo Forero Mendoza	Universidade Estadual de Rio de Janeiro, Brazil
Alvaro Gustavo Talavera	Universidad del Pacífico, Peru
Efraín Mayhua-López	Universidad Católica San Pablo, Peru
Yván Tupac	Universidad Católica San Pablo, Peru
Ana Teresa Tapia	Escuela Superior Politécnica del Litoral, Ecuador
Miguel Núñez del Prado	Universidad del Pacífico, Peru
Heitor Silvério Lopes	Universidade Tecnológica Federal de Paraná, Brazil
Waldimar Amaya	ICFO, Institute of Photonic Sciences, Spain
Leonardo Franco	Universidad de Málaga, Spain
Carlos Andrés Peña	University of Applied Sciences, Western Switzerland, Switzerland
Edwin Alexander Cerquera	University of Florida, USA
Nadia Nedjah	Universidade Estadual do Río de Janeiro, Brazil
María Daniela López de Luise	CI2S Lab, Argentina
Gustavo Eduardo Juárez	Universidad Nacional de Tucumán, Argentina

Contents

Engineering Applications

Biomedical Applications

A Transfer Learning Scheme for COVID-19 Diagnosis from Chest X-Ray Images Using Gradient-Weighted Class Activation Mapping

Ricardo Araguillin[1]([✉])[iD], Diego Maldonado[1][iD], Felipe Grijalva[2][iD],
Diego S. Benítez[2][iD], and Noel Pérez-Pérez[2][iD]

[1] Departamento de Automatización y Control Industrial,
Escuela Politécnica Nacional, Quito 170109, Ecuador
{ricardo.araguillin,diego.maldonadoa}@epn.edu.ec
[2] Colegio de Ciencias e Ingenierías, Universidad San Francisco de Quito USFQ,
Quito 170157, Ecuador
{fgrijalva,dbenitez,nperez}@usfq.edu.ec

Abstract. This paper proposes a new method for detecting COVID-19 in chest X-ray images by comparing it with existing models. Our proposed deep learning model is customized for an accurate diagnosis of COVID-19. To improve model performance and prevent overfitting, we first apply data augmentation techniques. Unlike traditional image segmentation methods, we use gradient-weighted class activation mapping (Grad-CAM) to highlight regions critical to identifying COVID-19. We then used transfer learning of Xception convolutional neural networks to extract the X-ray image data into a compact feature set. Finally, we design, parameterize, and train the neural classification network. The network showed impressive results, achieving an astonishing 97% accuracy in identifying healthy patients. At the same time, its detection rate in COVID-19-infected patients was 92%, making it a worthy competitor compared to other detection models.

Keywords: Xception · MobilNetV2 · deep learning · transfer learning · NNs · computer X-ray diagnostic tool · COVID-19

1 Introduction

COVID-19, caused by the SARS-CoV-2 virus, is a highly contagious viral respiratory illness. It has affected millions of people around the world, leading to respiratory symptoms. Governments and NGOs around the world have joined efforts to stop the spread of the virus. The World Health Organization (WHO) recommends early detection of positive patients for COVID-19. In this way, an early diagnosis and timely isolation are appropriate. Rapid detection tests such as

Supported by Escuela Politécnica Nacional.

reverse transcription polymerase chain reaction (RT-PCR) were initially costly but have since become more accessible. Another diagnostic method involves the analysis of lung images from tomography (CT) or chest radiographs (CXR), which, when properly processed, provide highly accurate results for medical personnel to identify COVID-19 positive patients [1,2]. This global health crisis quickly caught the attention of various scientific disciplines.

Governments and independent entities have conducted studies related to chest imaging, improving existing infrastructure, and incorporating artificial intelligence tools. Within the first studies, Nayak *et al.* proposed a Deep Learning (DL)-assisted automated method for the diagnosis of COVID-19. They evaluated eight pre-trained CNNs for the classification of infection from normal cases [3]. They develop comparative analyses among the models, considering factors such as the number of epochs, learning rate, batch size, and type of optimizers to find the best model.

Ozturk *et al.* created DarkCovidNet, an automated DL network for COVID-19 diagnosis based on X-ray images. It achieved 87.02% accuracy for multi-class (COVID-19, normal, and pneumonia) [4]. In related work, Wang and Wong designed COVID-Net, a deep CNN model with a testing accuracy of 92.6% [5]. Hemdan *et al.* developed COVIDX-Net, employing seven CNN models to train it. They achieved an accuracy of 87.02% [6]. Narin *et al.* obtained a testing accuracy of 98% using the ResNet50 model [7]. Sethy and Behera extracted features from various pre-trained CNN architectures using chest X-ray images, achieving the highest accuracy of 95.38% with ResNet50 coupled with Support Vector Machine (SVM) [8]. Ucar and Korkmaz [26] proposed COVIDiagnosis-Net, utilizing SqueezeNet and the Bayesian optimizer to achieve a testing accuracy of 98.3% in three-class classification cases [9]. Toğaçar *et al.* introduced an automated method to classify cases of COVID-19 from normal and pneumonia cases, employing MobileNetV2 and SqueezeNet with an SVM classifier. They reconstructed the original data set using fuzzy color and stacking techniques and further processed the features of the DL model using the Social Mimic Optimization (SMO) algorithm [10]. Recently, Farooq and Hafeez [28] developed COVIDResNet, a ResNet-based CNN model to classify COVID-19 and other cases (normal, bacterial pneumonia, viral pneumonia). They achieved an accuracy of 96.23% using a publicly available dataset (COVIDx), considering only 68 COVID-19 samples [11]. Chowdhury *et al.* [2] introduced a deep transfer learning (DTL) method using convolutional neural networks (CNN) for the automated detection of COVID-19 and viral pneumonia. Their study involved the training, validation, and testing of eight well-known CNN-based deep learning algorithms to distinguish between normal and pneumonia-afflicted patients using chest X-ray images..

Subsequently, Rahman *et al.* improved image processing by applying the gamma correction technique, which outperforms other techniques in detecting COVID-19, from simple and segmented lung CXR images [1]. In this case, the classification performance of plain CXR images is slightly better than that of segmented lung CXR images. In parallel, Polat *et al.* [12] explained the activation mapping method used to create activation maps that allow crucial radiograph areas to be highlighted to improve causality and intelligibility. The final nCOV-NET model optimized using the DenseNet-161 architecture for COVID-19 detection achieved a classification with 97.1% accuracy. In a more advanced way, Haq *et al.* [13] propose to improve network training through a data augmentation technique. For example, by rotating the existing images, new images are created that increase the practical training set of the R2DCNN model. Thus, they achieved an accuracy of 98%. Subsequently, studies propose the use of state-of-the-art deep Neural Network (NN) as feature extractors. In [14], a classification scheme based on CSEN (Convolution Support Estimation Network) was developed, with which CheXNet achieves more than 98% sensitivity and more than 95% specificity for the recognition of COVID-19.

Therefore, to obtain more manageable features for processing, it is necessary to reduce the number of image features while maintaining essential and relevant information. Some works propose improving the neural network's learning by having a segmentation stage. For example, some authors use convolutional networks for biomedical image segmentation such as U-Net, UNet++, VB-Net [15], and VNET-IR-RPN, ResNet-18 [16], to separate areas of interest. Thus, a machine learning algorithm can help health personnel diagnose and refer only those cases that require their expertise to the specialist doctor. In addition, properly using this algorithm can help reduce the staff's workload. Generally, the reviewed work does not consider the computational load demanded by the proposed models. However, in developing countries, such as Ecuador, there are rural areas where health personnel are limited and need specialists for a timely diagnosis. These areas generally need better technological infrastructure, so running a model like the one described above may be unfeasible. Therefore, developing a simple machine learning model with a low computational load and a high assertiveness rate would be beneficial.

In this sense, this paper proposes a methodology for developing a support tool aimed at facilitating the diagnosis of COVID-19. Initially, the database was balanced using data augmentation techniques. Instead of conventional image segmentation methods, this study suggests the use of Gradient-weighted Class Activation Mapping (Grad-CAM). Unlike image segmentation, which involves dividing the image into distinct regions, Grad-CAM focuses on highlighting the crucial regions directly related to the COVID-19 class. Subsequently, in the next stage, the feature vector of each image is reduced using CNN Xception to minimize computational training efforts. Then, an appropriate NN classifier, based on supervised learning, is designed and trained; this NN is responsible for identifying and learning the pattern of features in the images. Through this integrated approach, it becomes possible to achieve a high level of reliability in detecting the presence or absence of the disease. Lastly, a comparison of the proposed model in three different scenarios is carried out. This comparison demonstrates the performance of the model developed in this work.

This paper is organized as follows. Section 2 describes the data image set, its features, and processing issues. In Sect. 3 we describe the proposed methodology.

Section 4 presents the results obtained and a discussion of them. Finally, Sect. 5 presents the conclusions and possible future works.

2 Database

This Section discusses the data sources and discusses the limitations of traditional segmentation.

2.1 Database Description

Despite global COVID-19 cases, there are limited accessible chest X-ray databases. Therefore, the database 'Chest X-Ray Images', which is publicly available and was the winner of the Kaggle community COVID-19 dataset award, was used mainly. The dataset contains 21,000+ chest X-ray images (299 × 299 pixels, PNG format) available on Kaggle for academic use [17].

Our study used a subset of 13,808 images, divided into two groups: 10,192 healthy X-rays (negative class) from the Kaggle and Mendeley Dataset [18]. The second group consists of 3,616 lung images of patients infected with COVID-19, which is our positive class. From this group, 2473 images were taken from the PadChest dataset [19], 183 images from a medical school in Germany [20], 559 images from the Società Italiana di Radiologia Medica ed Interventistica (SIRM) [21], EURORAD [22], GitHub Chestxray [23] datasets, and Figshare repository [24] and 400 images from the GitHub COVID-CXNet dataset [25]. Figure 1 provides a visual representation of image sets and their sources.

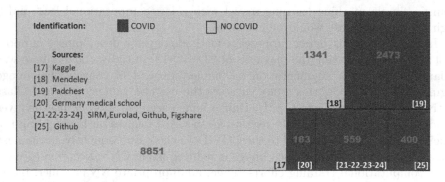

Fig. 1. Graphic representation of the set of images obtained according to their sources. Total number of images in the subset (in red) with their respective data source. (Color figure online)

2.2 Imagen Description

Each chest X-ray image is in RGB format and contains 268,203 features (299 × 299 × 3), which represents a large amount of information that must be processed per image, demanding a lot of computing power. Figure 2, shows four chest X-ray images: two from healthy patients and two from COVID-19 patients. Some regions aid COVID-19 diagnosis while others do not, making virus detection challenging. Typically, radiologists or pneumologists specializing in radiology or pneumology are required to determine whether the patient has the virus, in addition to other pathologies. It could be difficult for nonspecialist or inexperienced health personnel to identify whether or not the lungs have been infected with COVID-19.

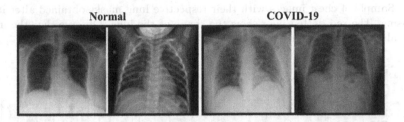

Fig. 2. Example of 4 chest x-ray images corresponding to healthy patients (two cases on the left) and those with COVID-19 (two cases on the right).

2.3 Image Segmentation Issues

The segmentation technique allows the extraction of only the specific area related to the disease. However, in many cases of patients positive for COVID-19, the segmented area corresponded to an area smaller than the lungs, resulting in a loss of information. Figure 3 shows four x-ray images with their respective masks obtained from the segmentation process, corresponding to two negative and two positive cases of COVID-19. Here, morphological operations were used to delimit the area of the lungs and create a mask. The dashed red line represents the area corresponding to the lung. In all cases, there is a tendency to lose information. This problem is especially evident in the right lung. As can be seen, there are differences at the edges that only a pulmonologist or a radiologist could interpret to decide if unimportant information is lost.

3 Proposed Methodology

The proposed machine learning strategy was divided into four stages: data augmentation, class activation mapping, feature extraction, and classification. Figure 4 represents the summary of the proposed methodology through sequential steps. Each of them is described below.

Data Augmentation. Since our database is unbalanced, as Fig. 1 shows, there is an approximate ratio of 3:1 between the negative and positive cases. Having data with such characteristics can cause overfitting problems in the machine learning model, as noted in previous works using neural networks for image

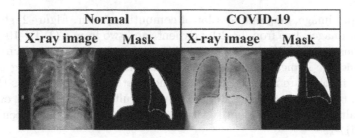

Fig. 3. Sample of chest images with their respective lung masks obtained after image processing. The red dotted lines show the shape of the lung. Observe that the masks do not always cover the entire lung area. (Color figure online)

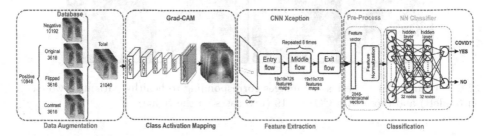

Fig. 4. Scheme of the machine learning model implemented for detecting COVID-19. The first block presents data augmentation techniques used to balance databases [26]. The second block represents the Grad-CAM process (Gradient-weighted Class Activation Mapping) [27,28]. The third block corresponds to the Xception convolutional neural network for dimensionality reduction [29,30]. The last block shows the neural network of the classifier composed of an input layer, an output layer, and two hidden layers [26]. The feature vectors are normalized to speed up the neural network's learning.

classification on unbalanced datasets [31,32]. For this reason, it was decided to increase the data by generating artificial images. To achieve this, flipping techniques and contrast change were implemented in all images corresponding to the positive class (i.e. COVID-19 positive patients). Crop, rotation and random Cop techniques were not considered because the lungs within the images have a defined spatial location. If they were moved, confusion would be caused in learning the CNN due to the change in their characteristics. With this, it was possible to increase the positive database for COVID-19 from 3616 to 10,848 images, which is slightly higher than the 10,192 images that comprise the database of people with a negative diagnosis, see Fig. 4. In this way, 21,040 images are used to learn and solve the unbalanced data problem.

Gradient-Weighted Class Activation Mapping (Grad-CAM). This method produces gradient-weighted heat maps, representing the activation classes on the images received at the input. This method is known as Grad-CAM and is performed using a pre-trained convolutional network to highlight

Normal	COVID-19
Original ⟶ CAP	Original ⟶ CAP

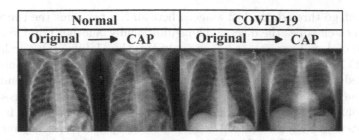

Fig. 5. Example of images obtained after Grad-CAM processing performed on images of healthy patients (left) and those with COVID-19 (right).

the main features of the images through different tones of color. With this, an activation class is associated with a specific output class and takes on a color [27,28]. These classes will allow one to indicate the importance of each pixel concerning the class by increasing or decreasing the intensity of the pixel. In this way, the most relevant features are highlighted in the output image, which in our case are the lungs. Figure 5 shows an example of four images that Grad-CAM has processed. A color gradient is observed over the lungs, highlighting features essential to this analysis. The result of this processing stage is 21,040 images with the new Grad-CAM format, each with its respective positive and negative diagnostic labels. The output of this stage becomes the input for feature extraction that will subsequently reduce the dimensions of the image to minimize the computational cost of learning.

Transfer Learning. This process seeks to create features derived from the original information that are informative and non-redundant, facilitating the next stage in the learning process. For this process, we have resorted to using the Xception convolutional neural network (CNN), a new deep convolutional neural network architecture inspired by Inception, where depth-separable convolutions have replaced Inception modules [29]. Xception has previously been trained with a set of 350 million images and 17000 classes. Given that the architecture of this neural network has the same number of parameters as Inception V3, experts consider that Xception has better performance due to the efficient use of model parameters [30]. The Xception architecture generally has 14 modules, each consisting of depth-wise separable convolution (DSC) and grouping layers. A DSC group is a combination of depth convolution and point convolution. These modules are grouped into three stages: Entry Flow, Middle Flow, and Exit Flow. In the first stage, the images of the $299 \times 299 \times 3$ characteristics are entered, and through four modules, the dimension of the data is transformed to $19 \times 19 \times 728$. These new data go to the second stage. The middle flow consists of 8 modules with no pooling layers, so the dimension of the data does not change. The data go through this stage eight times and then to the Exit Flow stage. In the last step, the data go through a 2-module that changes the dimension and converts it to 2048-dimensional feature vectors. As a result of this stage, the new images, represented by a vector of 2048 features, will go through the last stage, where an NN performs the learning process and then classifies the disease. [29]. Additionally, another notable approach in the field of deep learning for image analysis is the MobileNetV2 architecture. MobileNetV2, much like Xception, aims to enhance the extraction of informative and non-redundant features from original data, thereby facilitating subsequent stages in the learning process. This architecture builds upon the advances made in deep convolutional neural networks and represents a further evolution in the pursuit of efficient and effective model design. While Xception draws inspiration from the Inception model, MobileNetV2 incorporates innovative techniques and optimizations to achieve superior performance with a comparable number of parameters. Researchers have conducted extensive training, involving millions of images across thousands of classes, to refine MobileNetV2's ability to extract meaningful features. MobileNetV2 represents a promising advancement in deep learning architectures, offering researchers and practitioners a powerful tool for extracting valuable insights from complex image data.

Classification. The feature vectors obtained in the CNN are normalized before entering the classification neural network. According to the literature, normalizing the data accelerates the learning of the neural network [26]. Once the feature vectors were normalized, the entire dataset was divided into two subgroups, one for training (85%) and one for testing (15%). We separated 10% of the training set to be used as the validation set. The training data set used a five-fold cross-validation mechanism to evaluate the machine learning model implemented.

The neural network used for classification consists of four layers: the input layer with 2048 neurons, two hidden layers with 32 neurons each, and an output layer with 2 neurons to identify if the patient has COVID-19. The network was configured with a batch size of 16 and trained for 150 epochs. Figure 4 presents a scheme of the implemented model.

Accuracy was used to evaluate the model during the training stage. For the final performance evaluation, we calculated accuracy, recall, F1-score and precision, as well as the area under the ROC curve. The ROC curve plots the true positive rate (TPR) on the y-axis against the false positive rate (FPR) on the x-axis for different threshold decision values ranging from 0 to 1. We also calculated the confusion matrix using a five-fold cross-validation procedure and determined the overall accuracy in all runs.

4 Results

The results presented in this section were obtained after processing data from 21,040 X-ray images, where 10,848 correspond to positive patients and 10,192 to negative patients with COVID-19. This proposal is satisfactorily evaluated in a five-fold cross-validation process.

From the execution of the proposed five-fold cross-validation, a mean accuracy of 95.211% was obtained with a standard deviation (SD) of 0.364% (see Table 1). Furthermore, the best prediction model was found to achieve an accuracy of 95.739%, which was taken as a reference to carry out the final test and

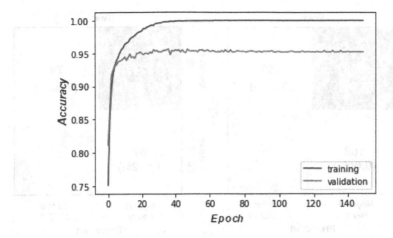

Fig. 6. Learning evolution of the proposed neural network when evaluated in each epoch.

Table 1. Accuracy-based performance for each fold

Fold	1	2	3	4	5
Accuracy (%)	94.804	95.526	95.100	95.739	94.887
Mean ± (SD) (%)	95.211±(0.364)				

present the results. Figure 6 illustrates the learning behavior of the best model (4th-Fold) with respect to the step during each epoch. The figure clearly shows how appropriate parameterization of the neural network aids the learning process in increasing its accuracy with each epoch.

Figure 7 shows the results of the confusion matrix for the training set (left) and the test set (right). Observe that the false negative rate increased in the test set. This might be due to the imbalanced distribution of the dataset.

Table 2 displays the model performance metrics for both training sets and test sets obtained from the confusion matrix (Fig. 7). We also measure the sensitivity and specificity of the model in detecting sick and healthy patients among all individuals.

During training, the model achieved a sensitivity of 98% and a specificity of 99%. However, during validation, we observed slightly lower results. The model was able to recognize healthy patients with 97% accuracy and identify patients with COVID-19 with 92% accuracy. The f1-scores were similar to the precision and sensitivity values in all cases. The difference between sensitivity and specificity could be attributed to an imbalance in the dataset.

Figure 8 displays the receiver operating characteristic (ROC) curve obtained for the test set. The curve clearly shows that the rate of true positives is higher than the rate of false negatives. The elbow of the curve indicates a balance point between sensitivity and specificity, which is approximately 0.9 for both cases. The area under the ROC curve reflects the quality of the model, and in this

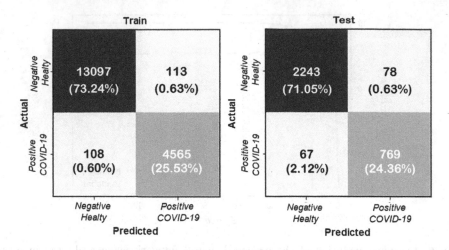

Fig. 7. Confusion matrix for the training (left) and test (right) sets, respectively. Negative cases represent healthy patients, and positive cases those with COVID-19.

Table 2. Metrics obtained from the confusion matrix.

Set	Class	Precision (%)	Recall (%)	F1-score (%)
Train	Negative	99	99	99
	Positive	98	98	98
Test	Negative	97	97	97
	Positive	91	92	91

case, the area under the curve (AUC) is 0.99, indicating that the implemented model is reliable, high-performance, and suitable for the task at hand.

4.1   Comparison with Other Models

In order to accurately assess the performance of our implemented model, we performed a comprehensive comparative analysis in the context of the existing scientific literature. The evaluation consists of two different comparisons: one with a model run on a similar dataset to ours, and another with a model run on a different dataset.

For the first comparison, we implemented a model with an architectural framework comparable to those examined in the literature. The architecture includes key components such as data augmentation, transfer learning using convolutional neural network (CNN) models, and classification schemes (see Fig. 9). In particular, Chowdur *et al.* and Nayak *et al.* performed a comprehensive comparison using eight different deep CNN models. Given the common use of the MobileNetV2 CNN architecture in the two aforementioned studies, we chose this specific architecture as the focus of direct comparison in our own experiments. This choice was made to provide a deeper and more meaningful evaluation of the performance of our model.

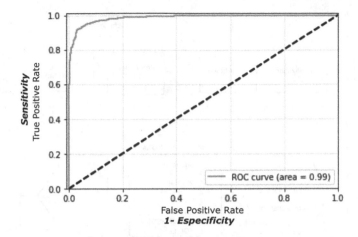

Fig. 8. ROC curve to evaluate the performance of the implemented machine learning model.

In order to evaluate the effectiveness of the method, a comprehensive series of tests was conducted to facilitate comparison and identify the best model. The results of these evaluations are shown in Table 3, which shows the results of five different training iterations in three different scenarios.

In Case A, the original dataset was transformed using the Xception architecture and then classified using a custom neural network. Case B, on the other hand, involves the use of data augmentation with a sequential combination of MobileNetV2, GradCam, and neural networks. Finally, Case C involves data augmentation and the sequential combination of Xception, GradCam, and neural networks. We carefully examine these three scenarios to provide a comprehensive overview of the performance of our approach.

Table 3. Comparison with MobileNet V2.

Configuration		Case A	Case B	Case C
		Original Database	Data augmentation	
		Xception	MobileNetV2	Xception
		NN	Gradcam	
			NN	
Fold	1	94,719	94,730	94,804
	2	95,654	93,382	95,260
	3	94,844	93,015	95,100
	4	95,910	93,137	95,739
	5	95,015	95,098	94,887
Mean (%)		*95,228*	*93,873*	*95,158*
SD (%)		*0,524*	*0,969*	*0,371*

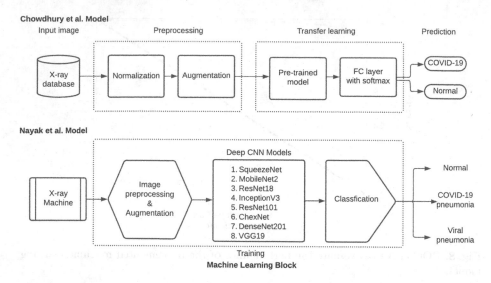

Fig. 9. Comparation of models based on Deep Learning Algothims and similar datbase.

To determine the differences between scenarios, we performed t-tests to compare the cases. The observations are as follows:

Case A vs. B: These cases show a statistically significant difference with a t-value of 2.7529 and a corresponding p-value of 0.0249. It is worth noting that including MobileNetV2 in the architecture results in a significant decrease in prediction performance.

Case A vs. C: On the contrary, there is no statistically significant difference between these cases, with a t-value of 0.2452 and a corresponding p-value of 0.8125. Although the introduction of the new architecture does not result in a significant improvement in accuracy, it does help reduce the standard deviation, thus improving the accuracy of the model predictions.

Case B vs. C: These cases also show a statistically significant difference with a t-value of −2.7712 and a p-value of 0.02424. This shows that incorporating Xception into the proposed architecture outperforms the MobileNetV2 implementation in terms of predictive performance.

Furthermore, in the course of comparing the outcomes achieved by our proposed model to those obtained by Chowdur and Nayak, a noteworthy observation emerges. The accuracy of our model is well within a predefined and deemed acceptable range, showcasing its competitiveness and efficacy in relation to the benchmark models established by Chowdur and Nayak (see Table 4). This alignment of performance underscores the robustness and viability of our proposed model within the context of the studied problem domain.

A further round of comparative analysis was conducted, this time juxtaposing our work with existing studies sourced from the literature. These previous studies encompassed divergent datasets and featured novel machine learning architectures, distinctly named after their respective authors. In particular, the

Table 4. Comparison with Chowdhur and Nayak models.

Schemes	Models	Accuracy [%]
Proposed Model	Xception	95.23
	MobileNetV2	93.87
Proposed by Chowdhury *et al.*	SqueezeNet	95.10
	MobileNetV2	96.22
	ResNet18	96.44
	InceptionV3	96.20
	ResNet101	96.22
	CheXNet	96.94
	DenseNet	97.94
	VGG19	96.00
Proposed by Nayak *et al.*	ResNet34	98.33
	ResNet50	97.50
	GoogleNet	96.67
	VGG-16	95.83
	AlexNet	97.50
	MobileNetV2	95.83
	InceptionV3	92.50
	SqueezeNet	96.67

diversity in the datasets ranged from a modest 50 images to a considerably larger dataset comprising 13,800 images.

Of particular interest, as elucidated in Table 5, is the remarkable performance of the models designed for relatively small datasets, consistently achieving accuracy levels that exceed the threshold 95%. This remarkable achievement can be attributed to the intricate and intricate architectures that underlie these models. On the contrary, when we scrutinize the model devised by Wang and Wong, which shares a database similar to the one utilized in our research, we find that its accuracy falls short of the Grad-CAM + Xception architecture presented in our study. This outcome implies that, despite the relative simplicity of our model's architecture, it demonstrates the capacity to attain a commendable level of accuracy in comparison to its peers.

Table 5. Comparison with models using other databases.

Autors	Deep Learning Model	Database Images	Images by Classes	Accurracy [%]
Nayak *et al.* [3]	ResNet-34	406	COVID-19: 203 Normal: 203	98.33
Ozturk *et al.* [4]	DarkCovidNet	1125	COVID-19: 125 Normal: 500 Pneumonia: 500	87.02
Wang and Wong [5]	COVID-Net	13800	COVID-19: 183 Normal: - Pneumonia: -	92.6
Hemdan *et al.* [6]	COVIDX-Net	50	COVID-19: 25 Normal: 25	90
Narin *et al.* [7]	ResNet-50	100	COVID-19: 50 Normal: 50	98
Sethy and Behera [8]	ResNet-50 and SVM	100	COVID-19: 50 Normal: 50	95.38
Ucar and Korkmaz [9]	Bayes-SqueezeNet	5949	COVID-19: 76 Normal: 1583 Pneumonia: 4290	98.3
Toğaçar *et al.* [10]	SqueezeNet and MobileNetV2 SMO and SVM	458	COVID-19: 295 Normal: 65 Pneumonia: 98	98.25
Farooq and Hafeez [11]	COVID-ResNet	5941	COVID-19: 68 Normal: - Bacterial Pneumonia: - Viral Pneumonia: -	96.23

5   Conclusions

This study proposes the development of a tool to diagnose COVID-19 by analyzing chest X-ray images. The proposed approach employs supervised learning techniques to analyze more than 21,000 images, leveraging data augmentation to improve model generalization and Gradient-weighted Class Activation Mapping (Grad-CAM) instead of conventional image segmentation techniques. Additionally, an Xception convolutional neural network is integrated into the transfer-learning process to reduce the image dimensions from $299 \times 299 \times 3$ to a vector of 2048 relevant features. Furthermore, a second neural network utilizes this information and trains itself to recognize the presence or absence of COVID-19. The results were validated using a five-fold cross-validation scheme, and the trained model was validated with approximately 18,000 images, achieving a prediction accuracy of 95.739%.

Future research could explore various topics, such as testing the proposed methodology with different neural network architectures, improving disease detection rates, using alternative data enhancement techniques, and testing new classification algorithms. Furthermore, there is the possibility of expanding the study to include other lung diseases and incorporating new chest radiographs into the training process.

Acknowledgements. The authors acknowledge support from Escuela Politécnica Nacional.

References

1. Rahman, T., et al.: Exploring the effect of image enhancement techniques on covid-19 detection using chest x-ray images. Comput. Biol. Med. **132**, 104319 (2021). https://www.ncbi.nlm.nih.gov/pmc/articles/PMC7946571/
2. Chowdhury, M.E., et al.: Can AI help in screening viral and covid-19 pneumonia? IEEE Access **8**, 132665–132676 (2020)
3. Nayak, S.R., Nayak, D.R., Sinha, U., Arora, V., Pachori, R.B.: Application of deep learning techniques for detection of covid-19 cases using chest x-ray images: A comprehensive study. Biomed. Signal Process. Control **64**, 102365 (2021)
4. Ozturk, T., Talo, M., Yildirim, E.A., Baloglu, U.B., Yildirim, O., Acharya, U.R.: Automated detection of covid-19 cases using deep neural networks with x-ray images. Comput. Biol. Med. **121**, 103792 (2020)
5. Wang, L., Lin, Z.Q., Wong, A.: Covid-net: a tailored deep convolutional neural network design for detection of covid-19 cases from chest x-ray images. Sci. Rep. **10**(1), 19549 (2020)
6. Hemdan, E.E.-D., Shouman, M.A., Karar, M.E.: Covidx-net: a framework of deep learning classifiers to diagnose covid-19 in x-ray images. arXiv preprint arXiv:2003.11055 (2020)
7. Narin, A., Kaya, C., Pamuk, Z.: Automatic detection of coronavirus disease (covid-19) using x-ray images and deep convolutional neural networks. Pattern Anal. Appl. **24**, 1207–1220 (2021)
8. Sethy, P.K., Behera, S.K.: Detection of coronavirus disease (covid-19) based on deep features (2020)
9. Ucar, F., Korkmaz, D.: Covidiagnosis-net: deep bayes-squeezenet based diagnosis of the coronavirus disease 2019 (covid-19) from x-ray images. Med. Hypotheses **140**, 109761 (2020)
10. Toğaçar, M., Ergen, B., Cömert, Z.: Covid-19 detection using deep learning models to exploit social mimic optimization and structured chest x-ray images using fuzzy color and stacking approaches. Comput. Biol. Med. **121**, 103805 (2020)
11. Farooq, M., Hafeez, A.: Covid-resnet: a deep learning framework for screening of covid19 from radiographs. arXiv preprint arXiv:2003.14395 (2020)
12. Polat, Ç., Karaman, O., Karaman, C., Korkmaz, G., Balci, M.C., Kelek, S.E.: Covid-19 diagnosis from chest x-ray images using transfer learning: enhanced performance by debiasing dataloader. J. X-Ray Sci. Technol. **29**, 19–36 (2021)
13. Haq, A.U., Li, J.P., Ahmad, S., Khan, S., Alshara, M.A., Alotaibi, R.M.: Diagnostic approach for accurate diagnosis of covid-19 employing deep learning and transfer learning techniques through chest x-ray images clinical data in e-healthcare. Sensors **21**, 8219 (2021)
14. Yamac, M., Ahishali, M., Degerli, A., Kiranyaz, S., Chowdhury, M.E.H., Gabbouj, M.: Convolutional sparse support estimator-based covid-19 recognition from x-ray images. IEEE Trans. Neural Netw. Learn. Syst. **32**, 1810–1820 (2021)
15. Shi, F., et al.: Review of artificial intelligence techniques in imaging data acquisition, segmentation, and diagnosis for covid-19. IEEE Rev. Biomed. Eng. **14**, 4–15 (2021). https://www.scopus.com/inward/record.uri?eid=2-s2.0-85083726857&doi=10.1109%2fRBME.2020.2987975&partnerID=40&md5=2b68188934d53b5f3a992cc2ca15d7e6

16. El Asnaoui, K., Chawki, Y.: Using x-ray images and deep learning for automated detection of coronavirus disease. J. Biomolec. Struct. Dyn. 1–12 (2020). https://www.scopus.com/inward/record.uri?eid=2-s2.0-85085621199&doi=10.10 80%2f07391102.2020.1767212&partnerID=40&md5=b3fdf5c116aede95befe0e2313 d5109b
17. Rsna pneumonia detection challenge — kaggle. https://www.kaggle.com/c/rsna-pneumonia-detection-challenge/data
18. Kermany, D., Zhang, K., Goldbaum, M.: Labeled optical coherence tomography (oct) and chest x-ray images for classification. Mendeley Data **2** (2018)
19. Bimcv-covid19 - bimcv. https://bimcv.cipf.es/bimcv-projects/bimcv-covid19/# 1590858128006-9e640421-6711
20. covid-19-image-repository/png at master ml-workgroup/covid-19-image-repository github. https://github.com/ml-workgroup/covid-19-image-repository/tree/maste r/png
21. Covid-19 database - sirm. https://sirm.org/category/senza-categoria/covid-19/
22. Homepage | eurorad. https://www.eurorad.org/
23. Github - ieee8023/covid-chestxray-dataset: we are building an open database of covid19 cases with chest xray or ct images. https://github.com/ieee8023/covid-chestxray-dataset
24. Arman, H., Majdabadi, M.M., Seokbum, K.: Covid19 chest xray image repository (2021). https://figshare.com/articles/dataset/COVID
25. Github - armiro/covid-cxnet: Covid-cxnet: Diagnosing covid-19 in frontal chest x-ray images using deep learning. Preprint available on arxiv: https://arxiv.org/abs/ 2006.13807. https://github.com/armiro/COVID-CXNet
26. Fenner, M.: Machine Learning with Python for Everyone. Addison-Wesley Professional, Boston (2019)
27. Selvaraju, R.R., Cogswell, M., Das, A., Vedantam, R., Parikh, D., Batra, D.: Grad-cam: visual explanations from deep networks via gradient-based localization. In: Proceedings of the IEEE International Conference on Computer Vision, pp. 618–626 (2017)
28. Panwar, H., Gupta, P., Siddiqui, M.K., Morales-Menendez, R., Bhardwaj, P., Singh, V.: A deep learning and grad-cam based color visualization approach for fast detection of covid-19 cases using chest x-ray and ct-scan images. Chaos Solitons and Fractals **140** (2020). https://www.scopus.com/inward/record.uri? eid=2-2.0-85091328347&doi=10.1016%2fj.chaos.2020.110190&partnerID=40& md5=0b414b46ab1eecdccbb8a410cac9e172
29. Chollet, F.: Xception: deep learning with depthwise separable convolutions. In: Proceedings of the IEEE Conference on Computer Vision and Pattern Recognition, pp. 1251–1258 (2017)
30. Sharma, S., Kumar, S.: The xception model: a potential feature extractor in breast cancer histology images classification. ICT Exp. **8**, 101–108 (2022)
31. Rivera, D., Grijalva, F., Acurio, B.A.A., Álvarez, R.: Towards a mobile and fast melanoma detection system. In: 2019 IEEE Latin American Conference on Computational Intelligence (LA-CCI), pp. 1–6. IEEE (2019)
32. Parra, C., Grijalva, F., Núñez, B., Núñez, A., Pérez, N., Benítez, D.: Automatic identification of intestinal parasites in reptiles using microscopic stool images and convolutional neural networks. PLoS ONE **17**(8), e0271529 (2022)

On the Comparison of Multilayer Perceptron and Extreme Learning Machine for Pedaling Recognition Using EEG

Cristian Felipe Blanco-Díaz[1]([✉])[iD], Cristian David Guerrero-Mendez[1][iD],
Teodiano Bastos-Filho[1][iD], Andrés Felipe Ruiz-Olaya[2][iD],
and Sebastián Jaramillo-Isaza[3][iD]

[1] Postgraduate Program in Electrical Engineering, Federal University of Espirito
Santo (UFES), 29075-910 Vitória, Brazil
{cblanco88,crguerrero69}@uan.edu.co, teodiano.bastos@ufes.br
[2] Faculty of Mechanical, Electronic and Biomedical Engineering,
Antonio Nariño University (UAN), Cra. 3 E No 47A 15, Bogotá, Colombia
andresru@uan.edu.co
[3] Universidad del Rosario, School of Medicine and Health Sciences Bogota,
Bogotá, Colombia

Abstract. Brain-Computer Interfaces (BCIs) have gained significant attention in recent years for their role in connecting individuals with external devices using neural signals. Electroencephalography (EEG)-based BCIs, in combination with Motorized Mini Exercise Bikes (MMEBs), have emerged as promising tools for post-stroke patient rehabilitation. Nevertheless, the EEG signal-to-noise ratio (SNR) remains a challenge, susceptible to interference from physical and mental artifacts, thereby compromising the accuracy of motor task recognition, such as pedaling. This limitation hampers the effectiveness of lower-limb rehabilitation devices. In this study, we propose a comparative study which uses Multilayer Perceptron (MLP) and Extreme Learning Machine (ELM) to accurately identify from EEG signals when a subject is engaged in pedaling tasks. The results outperform those reported in the literature, achieving a remarkable Accuracy of 0.97 and a negligible False Positive Rate close to zero, resulting in an overall performance of 0.77 and 0.24, respectively. Additionally, we conducted an evaluation of four distinct frequency bands during the filtering process, with the most promising outcomes achieved within the 3 to 7 Hz frequency band. These findings support the conclusion that our proposed methodology is well-suited for the real-time detection of lower-limb tasks using EEG signals, thus offering potential applications in the control of robotic BCIs for rehabilitation purposes.

Keywords: Brain-Computer Interfaces · Lower-limb recognition · Multilayer Perceptron · Extreme Learning Machine · Control for rehabilitation devices

A. D. Orjuela-Cañón et al. (Eds.): ColCACI 2023, CCIS 1865, pp. 19–29, 2024.
https://doi.org/10.1007/978-3-031-48415-5_2

1   Introduction

Brain-Computer Interface (BCI) systems have emerged as transformative tools, enabling individuals to communicate with their environment through computer interfaces without the need for direct engagement with the Peripheral Nervous System (PNS) while harnessing information from the Central Nervous System (CNS) [22]. Among the myriad technologies available for BCI development, Electroencephalography (EEG) has emerged as a particularly valuable asset. EEG allows for the non-invasive monitoring of brain electrical activity through electrodes, making it possible to discern patterns related to a user's intentions, which can subsequently be translated into control commands for robotic devices. As a result, EEG-based BCIs have found applications in controlling prostheses [14], exoskeletons [2], walking aids [8], and, most notably, in the rehabilitation of post-stroke patients through their integration with Motorized Mini-Exercise Bikes (MMEBs) [6,18,21].

Prior research has placed significant emphasis on the detection and imaging of pedaling movements using EEG, given its potential to yield restorative benefits in Activities of Daily Living (ADLs) associated with lower-limb cyclic movements, such as walking [23]. However, it is important to acknowledge that EEG signals often suffer from a low Signal-to-Noise Ratio (SNR) and are susceptible to artifacts induced by various physical and mental factors [1,5,14]. Consequently, the development of methodologies that utilize EEG for such purposes poses a formidable challenge to the BCI research community.

In response to this challenge, several researchers have introduced novel methodologies for the identification of motor tasks through EEG. For example, Liu et al. proposed a method for classifying ankle movements based on delta-band features [12]. Rodriguez-Ugarte et al. explored frequency band features between 6 and 30 Hz for classifying pedal Motor Imagery (MI) [16,17]. Similarly, Romero et al. leveraged methods rooted in Riemannian Geometry to identify lower-limb Motor Imagery within the 0.1 to 40 Hz frequency band [18]. Despite significant progress in lower-limb MI identification, a notable gap persists in the classification of executed movements, a challenge exacerbated by the artifacts introduced by limb movements [14]. Blanco-Díaz et al. conducted a study reconstructing kinematic data during pedaling using EEG features extracted from the delta band (0.5–4 Hz) [4]. Conversely, Storzer et al. investigated pedaling tasks within the beta frequency band (23–30 Hz) [20], although classification techniques for pedaling task recognition were not explored in these studies.

Artificial Neural Networks (ANNs) have emerged as a promising avenue for improving the detection of motor tasks via EEG. The Multilayer Perceptron (MLP), in particular, has demonstrated exceptional capabilities in classifying both upper and lower-limb-related tasks, thanks to its deep learning-based classification prowess [13]. Conversely, the Extreme Learning Machine (ELM) has shown promise in MI identification, outperforming traditional classifiers such as Linear Discriminant Analysis (LDA) or Support Vector Machine (SVM) [11,19]. Notably, ELM has also been employed for identifying pedaling motor tasks [3], although a comprehensive comparison with other configurations, such as MLP-based methods, remains unexplored.

In light of the aforementioned research landscape, the full potential of Artificial Neural Networks in identifying lower-limb tasks associated with pedaling remains largely untapped. Consequently, this study aims to enhance the detection of lower-limb tasks during pedaling using EEG. To achieve this objective, we developed an algorithmic approach based on temporal EEG signal feature extraction, employing both MLP and ELM to discriminate between resting and pedaling states. Furthermore, recognizing the absence of a widely accepted reference standard in the literature for identifying pedaling tasks, we conducted an extensive evaluation of four distinct frequency bands to ascertain the most suitable one for discriminating this crucial motor information.

2 Methodology

The methodology was segmented as follows: EEG acquisition, signal preprocessing, feature extraction, and classification using MLP and ELM. A summary of the methodology is presented in Fig. 1.

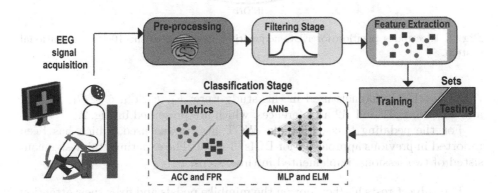

Fig. 1. Block diagram of the methodology for the identification of pedaling tasks using EEG. ANN corresponds to the MLP and ELM.

2.1 EEG Dataset

An in-house database was acquired for motor pedaling tasks. The signals were measured with an OpenBCI board of eight channels according to the 10–20 international system and OpenVibe software at 250 Hz as the sampling frequency. EEG channels were selected based on the primary region where Event Related Desynchronization/Synchronization (ERD/ERS) occurs during lower-limb movement for cycling tasks, which is located in the parietal-central cortex

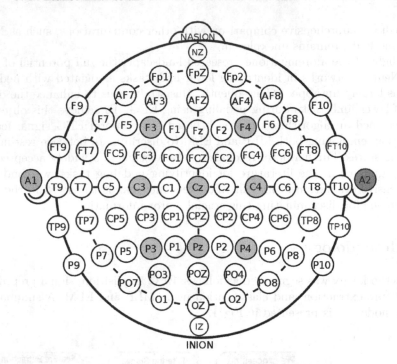

Fig. 2. Electrodes position for EEG acquisition according to the 10–20 international system.

[3,6]. The EEG channels used in this study were: F_3, F_4, C_3, C_Z, C_4, P_3, P_Z, and P_4, using $A1$ and $A2$ as references, which are presented in Fig. 2.

For the pedaling tasks, Minibike WCT fitness was used, which has been reported in previous approaches for BCIs [3,4,18]. The experimental design consisted of two sessions, implemented in three steps.

- The subject rests his/her foot on the minibike pedals and fixes their attention on a black screen for 5 s without engaging in any mental activity.
- A green cross is displayed on the screen, which is the instruction for pedaling the minibike during 5 s.
- The subject rests between 5 or 7 s, and the cycle is repeated 30 times (trials).

An adequate number of participants was determined based on a statistical power analysis performed using G*power software [7]. This analysis was performed to verify whether the number of subjects used was sufficient to generalize the model based on literature results. To address this, a G*power $d = 2.13$, a value of $\alpha = 0.05$, and a power of 0.90 were used, as reported in [3,10]. The minimum number of subjects calculated using the G* Power software for the Wilcoxon signed-rank test was equal to four; however, in this study, a deviation of 20% was considered.

Five healthy subjects (aged between 25 and 32 years; three male and two female) voluntarily participated in this study. The experiment was conducted in accordance with the Declaration of Helsinki and the rules of the Ethics Committee of the Federal University of Espirito Santo (UFES/Brazil) under the number CAAE:39410614.6.0000.5060.

2.2 Pre-processing

The signals were segmented in a time window between 1 and 4.5 s after the trigger, considering that several studies have reported that during this time, motor tasks are more discriminant [1,9]. A Common Average Reference (CAR) filter was implemented to reduce common noise. The signals were then filtered using a zero-phase 4th-order bandpass Butterworth filter between 3 and 30 Hz [3, 4,6]. As an additional step, a filter bank was implemented, highlighting the lack of literature regarding the evaluation of EEG spectral behavior during pedaling. Therefore, the signals were filtered using four different Frequency Bands (FB): **FB1:** 3–7 Hz, **FB2:** 7–13, **FB3:** 7–30 Hz, and **FB4:** 3–30 Hz.

2.3 Feature Extraction

Two features widely reported in the literature were used for EEG classification based on the Root Mean Square (RMS) and Power Spectral Density (PSD) [1,14,15]. These techniques are represented by the following Eqs. 1 and 2.

$$RMS = \sqrt{\frac{1}{N}\sum_{n=1}^{N} X[n]^2}, \tag{1}$$

$$PSD = \sum_{n=1}^{N}(P(X[n]))^2, \tag{2}$$

where RMS is the RMS value of the signal, N is the number of samples in the time window, $X[n]$ corresponds to the segmented EEG signal after the filtering stage according to the FB, PSD is the signal PSD, and $P(X[n])$ corresponds to the Welch Power after the filtering stage according to the FB.

2.4 Multilayer Perceptron (MLP)

The Multilayer Perceptron (MLP) is a well-established type of Artificial Neural Network (ANN) widely employed in the field of machine learning. It comprises multiple layers of interconnected neurons, including an input layer, one or more hidden layers, and an output layer. Within this architecture, each neuron in a given layer is connected to every neuron in the subsequent layer, facilitating the seamless flow of information throughout the network. This study employed a hidden layer MLP, which was trained using the Levenberg-Marquardt algorithm. This algorithmic choice was made to enhance the network's capacity for learning and its ability to model complex relationships within the data.

2.5 Extreme Learning Machine (ELM)

An ELM is a neural network family of Single Hidden Layer Feedback Neural Networks (SHLFN) [3]. It is composed of a single hidden layer with its respective activation function, where the objective is to train the network by modifying the weights of the neurons in the hidden layer to such an extent that it allows the classification of a group of input variables [19]. ELM has advantages over other types of classifiers considering that its training time is short [3].

2.6 Training and Testing

To generalize the classification models, a k-fold cross-validation methodology was implemented, where the total feature dataset was separated into two parts (training and validation) with a random number of k times. This process was repeated for $k = 5$ and the validation metrics were averaged. This process is performed for the features extracted from each frequency band configuration (FB1, FB2, FB3, and FB4). To find the most suitable configuration of hidden neurons for the two ANNs, the number of neurons were adjusted and validated for 10 values, that is: 10, 20, 30, 40, 50, 60, 70, 80, 90, 100 using information from FB1.

2.7 Evaluation Metrics

To evaluate the models performance, the classification Accuracy (ACC), and False Positive Rate (FPR) were computed. The performance metrics were defined by Eqs. 3 and 4.

$$ACC = \frac{TP + TN}{TP + TN + FP + FN}, \tag{3}$$

$$FPR = \frac{FP}{TN + FP}, \tag{4}$$

where TP, TN, FP, and FN correspond to the true positives, true negatives, false positives, and false negatives, respectively.

3 Results and Discussion

The optimization results of the ANNs are shown in Fig. 3, which shows the variation in the Acc and FPR metrics varying the number of neurons in the hidden layers. From this Figure, it can be noted that a higher number of neurons in the hidden layer does not result in better performance. For instance, the best configuration for the MLP was with 30 hidden neurons, which represented an ACC of 0.76±0.08, and an FPR of 0.28±0.17. Regarding the ELM, the highest ACC (0.72±0.15) and FPR (0.24±0.21 were obtained with 10 hidden neurons. Therefore, these configurations in the hidden layer were selected to post-processing of each ANN.

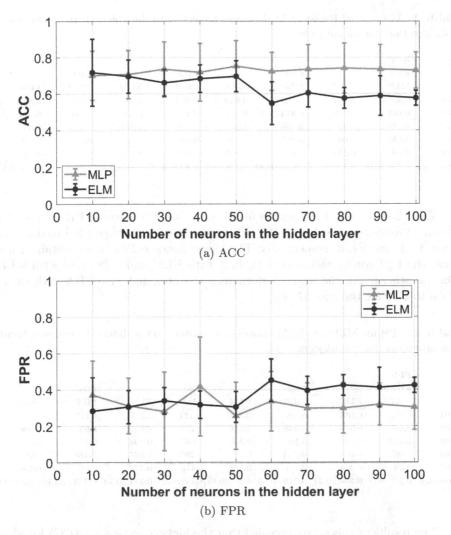

Fig. 3. ACC and FPR for the different number of neurons in hidden layer for the MLP and the ELM.

Subsequently, four frequency bands are evaluated, and the results for the ACCs are listed in Table 1. The highest overall performance was obtained with the FB1 configuration with an ACC of 0.77±0.08, with MLP, and 0.72±0.15, with ELM. The lowest performance was obtained with the FB2 configuration with $ACCs$ of 0.61±0.07 and 0.60±0.13 with MLP and ELM, respectively. Subject S05 obtained the highest average performance with $ACCs$ of 0.88 and 0.97 using the FB1 configuration with the MLP and ELM, respectively. Note that in the proposed model, ACC and FPR are inverse variables; thus, the appropriate classification model is obtained with ACC close to 1 and FPR close to 0.

Table 1. ACC for MLP and ELM classifiers according to the different frequency bands configurations for all subjects.

	ACC							
	FB1		FB2		FB3		FB4	
	MLP	ELM	MLP	ELM	MLP	ELM	MLP	ELM
S01	0.667	**0.833**	0.611	**0.639**	**0.861**	0.833	0.833	**0.861**
S02	**0.667**	0.514	**0.514**	0.429	0.486	0.486	0.514	**0.571**
S03	**0.750**	0.667	**0.556**	0.500	0.444	**0.500**	0.639	**0.694**
S04	**0.857**	0.600	**0.657**	0.571	**0.800**	0.629	**0.686**	0.514
S05	0.886	**0.971**	0.714	**0.857**	0.771	0.771	0.743	**0.771**
Average	**0.765±0.08**	0.717±0.15	**0.610±0.07**	0.599±0.13	**0.673±0.16**	0.644±0.13	**0.683±0.10**	0.683±0.12

Table 2 shows the *FPR* results for all subjects. The lowest FPR values were obtained for Subject S04 with the FB1 configuration, corresponded to 0.17 and 0 for MLP and ELM, respectively. The lowest average *FPR* value obtained was with the FB1 configuration of 0.24±0.21 with ELM and 0.28±0.18 with MLP. On the other hand, the worst performance was obtained with FB2, with values close to 0.28 for the two ANNs.

Table 2. FPR for MLP and ELM classifiers according to the different frequency bands configurations for all subjects.

	FPR							
	FB1		FB2		FB3		FB4	
	MLP	ELM	MLP	ELM	MLP	ELM	MLP	ELM
S01	0.214	**0.071**	0.316	**0.316**	**0.177**	0.235	0.177	0.177
S02	0.667	0.667	**0.188**	0.313	0.750	**0.600**	0.600	**0.400**
S03	0.158	0.263	0.308	0.308	0.667	**0.444**	0.421	**0.316**
S04	0.167	**0.000**	0.333	0.389	**0.267**	0.400	**0.188**	0.688
S05	0.200	0.200	0.235	**0.059**	0.412	**0.177**	0.273	**0.136**
Average	0.281±0.18	**0.240±0.21**	**0.276±0.05**	0.277±0.10	0.454±0.20	**0.371±0.14**	**0.332±0.15**	0.343±0.18

The results of this study revealed that the highest accuracy (*ACC*) for identifying pedaling tasks through EEG signals with Multilayer Perceptron (MLP) was achieved with FB1, while the best False Positive Rate (FPR) performance was attained with FB1 when utilizing the Extreme Learning Machine (ELM). These findings align with existing literature, which suggests that the most discriminative information for lower-limb movement execution can be gleaned from the frequency range of 3–7 Hz [3,4,6]. This study represents an advancement over previous research, as many studies on pedaling tasks have predominantly employed frequency bands spanning 7–30 Hz or 3–30 Hz (FB3 and FB4, respectively) [6,18,21]. This implies that selective filtering within specific frequency bands can indeed enhance the performance of Brain-Computer Interfaces (BCIs), as corroborated by prior investigations [3].

When examining the results on a per-subject basis, Subject S05 exhibited outstanding performance, achieving a maximum *ACC* of 0.97, while the lowest

FPR was recorded for Subject S04, approaching zero. Conversely, the poorest performance was observed in the data corresponding to Subject S03, with an ACC of 0.44, and Subject S02 for FPR. However, these outcomes were obtained using the FB3 frequency band.

In comparison to earlier literature, this study achieved overall maximum $ACCs$ of 0.77 and 0.72 and respective peaks of 0.86 and 0.97 using MLP and ELM. These results outperform existing techniques for lower-limb recognition. For instance, Ugarte et al. reported an ACC of 0.61 [16], whereas Liu et al., in a study focused on identifying ankle movements, achieved an ACC exceeding 0.81 [12]. It's important to note that these studies involve different types of movements, making direct comparisons challenging. Some studies on pedaling Motor Imagery (MI) tasks have reported ACC values of approximately 0.86 or FPR values of around 0.19, employing Riemannian computational strategies [18]. Nonetheless, it's essential to emphasize that these MI-based strategies may not be suitable for lower-level controller design due to their discrete system activation approach [4]. Furthermore, pedaling movements explored through various spectral analyses, such as filter banks, have not been extensively investigated in the literature [6,18]. Consequently, the utilization of Artificial Neural Networks (ANNs) like MLP and ELM for lower-limb task classification remains relatively unexplored.

Blanco-Díaz et al. proposed a first approximation to the ELM to classify pedaling tasks, obtaining ACC close to 0.73 and FPR close to 0.28 [3]. However, a limitation of their study is the absence of a comparative analysis between ELM and other classifiers. For this reason, the contributions of the proposed methodology are of significance, where it was possible to improve the average performance metrics in comparison to literature, enhancing by 5% the value of ACC using MLP and 4% for FPR using ELM compared with the results reported by [3].

Furthermore, it is noteworthy that MLP generally exhibited superior performance, albeit with a hidden layer comprising a higher number of neurons than ELM. This discrepancy may raise concerns regarding real-time usability due to extended training times. Consequently, it is advisable to conduct further investigations into optimizing both configurations to design more complex pedaling task recognition systems, which would ultimately lead to the development of more robust lower-limb rehabilitation systems.

4 Conclusion

The results of this study prove the feasibility of employing Root Mean Square (RMS) and Power Spectral Density (PSD) techniques in conjunction with Multilayer Perceptron (MLP) and Extreme Learning Machine (ELM) for the classification of lower-limb tasks related to pedaling. A remarkable maximum accuracy (ACC) of 0.97 was achieved for a single subject, with average ACC and FPR values standing at 0.77 and 0.24, respectively. Moreover, the investigation identified a discriminative frequency band for this classification task, with the most favorable outcomes occurring within the 3–7 Hz frequency band. This observation underscores the potential for the adoption of more intricate computational methods involving filter banks in conjunction with strategies based on Artificial Neural Networks (ANNs). In essence, this study affirms the precision of the proposed methodology in identifying pedaling tasks through EEG signals, thereby paving the way for its practical implementation in real-time Brain-Computer Interface (BCI) systems.

Future research activities will be directed toward the development of a system capable of facilitating Motorized Mini-Exercise Bikes (MMEBs) to assist with pedaling. The overarching goal is to enhance lower limb rehabilitation, particularly for post-stroke patients, by harnessing the insights gained from this study.

Acknowledgment. The authors would like to thank the Federal University of Espírito Santo (UFES) for the support to this research, and FAPES/I2CA (Resolution N° 285/2021) by the Master's scholarships awarded to the first two authors.

References

1. Abiri, R., Borhani, S., Sellers, E.W., Jiang, Y., Zhao, X.: A comprehensive review of EEG-based brain-computer interface paradigms. J. Neural Eng. **16**(1), 011001 (2019). https://doi.org/10.1088/1741-2552/aaf12e
2. Biao, L., Youwei, L., Xiaoming, X., Haoyi, W., Longhan, X.: Design and control of a flexible exoskeleton to generate a natural full gait for lower-limb rehabilitation. J. Mech. Robot. **15**(1), 011005 (2022). https://doi.org/10.1115/1.4054248
3. Blanco-Díaz, C., Guerrero-Méndez, C., Bastos-Filho, T., Ruiz-Olaya, A., Jaramillo-Isaza, S.: Detection of pedaling tasks through EEG using extreme learning machine for lower-limb rehabilitation brain-computer interfaces. In: 2023 IEEE Colombian Conference on Applications of Computational Intelligence (ColCACI), pp. 1–5 (2023). https://doi.org/10.1109/ColCACI59285.2023.10225911
4. Blanco-Díaz, C.F., Guerrero-Mendez, C.D., Delisle-Rodriguez, D., de Souza, A.F., Badue, C., Bastos-Filho, T.F.: Lower-limb kinematic reconstruction during pedaling tasks from EEG signals using unscented Kalman filter. In: Computer Methods in Biomechanics and Biomedical Engineering , pp. 1–11 (2023). https://doi.org/10.1080/10255842.2023.2207705
5. Blanco-Díaz, C.F., Guerrero-Méndez, C.D., Bastos-Filho, T., Jaramillo-Isaza, S., Ruiz-Olaya, A.F.: Effects of the concentration level, eye fatigue and coffee consumption on the performance of a BCI system based on visual ERP-p300. J. Neurosci. Methods. **382**, 109722 (2022). https://doi.org/10.1016/j.jneumeth.2022.109722, https://www.sciencedirect.com/science/article/pii/S0165027022002485
6. Delisle-Rodriguez, D., et al.: System based on subject-specific bands to recognize pedaling motor imagery: towards a BCI for lower-limb rehabilitation. J. Neural Eng. **16**(5), 056005 (2019)
7. Faul, F., Erdfelder, E., Lang, A.G., Buchner, A.: G* power 3: a flexible statistical power analysis program for the social, behavioral, and biomedical sciences. Behav. Res. Methods **39**(2), 175–191 (2007)
8. Gelaw, A.Y., Janakiraman, B., Teshome, A., Ravichandran, H.: Effectiveness of treadmill assisted gait training in stroke survivors: a systematic review and meta-analysis. Global Epidemiol. **1**, 100012 (2019)

9. Guerrero-Mendez, C.D., et al.: EEG motor imagery classification using deep learning approaches in Naïve BCI users. Biomed. Phys. Eng. Express **9**(4), 045029 (2023). https://doi.org/10.1088/2057-1976/acde82

10. Hosseini, S.M., Shalchyan, V.: State-based decoding of continuous hand movements using EEG signals. IEEE Access (2023)

11. Li, J., Li, Y., Du, M.: Comparative study of EEG motor imagery classification based on DSCNN and elm. Biomed. Signal Process. Control. **84**, 104750 (2023). https://doi.org/10.1016/j.bspc.2023.104750, https://www.sciencedirect.com/science/article/pii/S1746809423001830

12. Liu, D., et al.: EEG-based lower-limb movement onset decoding: continuous classification and asynchronous detection. IEEE Trans. Neural Syst. Rehabil. Eng. **26**(8), 1626–1635 (2018). https://doi.org/10.1109/TNSRE.2018.2855053

13. Narayan, Y.: Analysis of MLP and DSLVQ classifiers for EEG signals based movements identification. In: 2021 2nd Global Conference for Advancement in Technology (GCAT), pp. 1–6. IEEE (2021)

14. Padfield, N., Zabalza, J., Zhao, H., Masero, V., Ren, J.: Eeg-based brain-computer interfaces using motor-imagery: techniques and challenges. Sensors **19**(6), 1423 (2019). https://doi.org/10.3390/s19061423, https://www.mdpi.com/1424-8220/19/6/1423

15. Phinyomark, A., Phukpattaranont, P., Limsakul, C.: Feature reduction and selection for EMG signal classification. Expert Syst. App. **39**(8), 7420–7431 (2012). https://doi.org/10.1016/j.eswa.2012.01.102, https://www.sciencedirect.com/science/article/pii/S0957417412001200

16. Rodríguez-Ugarte, M., Angulo-Sherman, I.N., Iáñez, E., Ortiz, M., Azorín, J.M.: Preliminary study of pedaling motor imagery classification based on EEG signals. In: 2017 International Symposium on Wearable Robotics and Rehabilitation (WeRob), pp. 1–2 (2017). https://doi.org/10.1109/WEROB.2017.8383851

17. Rodríguez-Ugarte, M., Iáñez, E., Ortíz, M., Azorín, J.M.: Personalized offline and pseudo-online BCI models to detect pedaling intent. Front. Neuroinform. **11**, 45 (2017). https://doi.org/10.3389/fninf.2017.00045, https://www.frontiersin.org/articles/10.3389/fninf.2017.00045

18. Romero-Laiseca, M.A., et al.: A low-cost lower-limb brain-machine interface triggered by pedaling motor imagery for post-stroke patients rehabilitation. IEEE Trans. Neural Syst. Rehabil. Eng. **28**(4), 988–996 (2020)

19. She, Q., Zou, J., Luo, Z., Nguyen, T., Li, R., Zhang, Y.: Multi-class motor imagery EEG classification using collaborative representation-based semi-supervised extreme learning machine. Med. Biol. Eng. Comput. **58**, 2119–2130 (2020)

20. Storzer, L., et al.: Bicycling and walking are associated with different cortical oscillatory dynamics. Front. Hum. Neurosci. **10**, 61 (2016)

21. Villa-Parra, A.C., et al.: Control of a robotic knee exoskeleton for assistance and rehabilitation based on motion intention from SEMG. Res. Biomed. Eng. **34**, 198–210 (2018)

22. Wolpaw, J.R., Birbaumer, N., McFarland, D.J., Pfurtscheller, G., Vaughan, T.M.: Brain-computer interfaces for communication and control. Clin. Neurophysiol. **113**(6), 767–791 (2002). https://doi.org/10.1016/S1388-2457(02)00057-3

23. Yuan, Z., et al.: Effect of BCI-controlled pedaling training system with multiple modalities of feedback on motor and cognitive function rehabilitation of early subacute stroke patients. IEEE Trans. Neural Syst. Rehabil. Eng. **29**, 2569–2577 (2021)

Self-supervised Deep-Learning Segmentation of Corneal Endothelium Specular Microscopy Images

Sergio Sanchez[1]([✉]), Kevin Mendoza[1], Fernando Quintero[1],
Angelica M. Prada[2,3,4], Alejandro Tello[2,3,4,5], Virgilio Galvis[2,3,4],
Lenny A. Romero[6], and Andres G. Marrugo[1]

[1] Facultad de Ingeniería, Universidad Tecnológica de Bolívar, Cartagena, Colombia
ssanchez@utb.edu.co
[2] Centro Oftalmológico Virgilio Galvis, Floridablanca, Colombia
[3] Fundación Oftalmológica de Santander FOSCAL, Floridablanca, Colombia
[4] Facultad de Salud, Universidad Autónoma de Bucaramanga UNAB,
Bucaramanga, Colombia
[5] Facultad de Salud, Universidad Industrial de Santander UIS,
Bucaramanga, Colombia
[6] Facultad de Ciencias Básicas, Universidad Tecnológica de Bolívar,
Cartagena, Colombia

Abstract. Computerized medical evaluation of the corneal endothelium is challenging because it requires costly equipment and specialized personnel, not to mention that conventional techniques require manual annotations that are difficult to acquire. This study aims to obtain reliable segmentations without requiring large data sets labeled by expert personnel. To address this problem, we use the Barlow Twins approach to pre-train the encoder of a UNet model in an unsupervised manner. Then, with few labeled data, we train the segmentation. Encouraging results show that it is possible to address the challenge of limited data availability using self-supervised learning. This model achieved a precision of 86%, obtaining a satisfactory performance. Using many images to learn good representations and a few labeled images to learn the semantic segmentation task is feasible.

Keywords: Self-supervised · deep learning · segmentation · corneal endothelium

1 Introduction

The corneal endothelium is a set of hexagonal cells of vital importance for maintaining the transparency of the cornea. Which progressively deteriorates as age increases. This loss can be aggravated by surgical trauma or certain diseases such as Fuchs' corneal dystrophy [1,2].

Eye problems are highly prevalent and remain untreated in many cases. Among these diseases, Fuchs' corneal endothelial dystrophy affects the corneal

A. D. Orjuela-Cañón et al. (Eds.): ColCACI 2023, CCIS 1865, pp. 30–42, 2024.
https://doi.org/10.1007/978-3-031-48415-5_3

endothelium, which is responsible for corneal transparency [3]. According to WHO reports, greater vigilance is needed to guarantee attention to the ophthalmological needs of communities to improve prevention, early detection, treatment, and rehabilitation [4].

Among the challenges of studying the cornea is Fuchs' dystrophy, which is characterized by the accumulation of fluid in the cornea, located in the front part of the eye. This causes the cornea to swell and become thicker. This condition manifests itself with blurred vision and eye discomfort. Causing loss of visual acuity. This pathology of the endothelium is produced in the deepest layer of the cornea, where the cells responsible for maintaining corneal transparency are located. This disease can be caused by drug use, aging, surgeries and inflammation.

Today, there are different technological tools for assessing the state of the corneal endothelium through morphometric parameters, such as cell density. The most used device is the specular microscope. However, several studies have shown that conventional approaches for estimating endothelial morphometric parameters fail in the presence of endotheliopathies such as Fuchs' CE distrophy [5–8].

Recent deep learning-based approaches have significantly improved image segmentation of corneal endothelium and estimation of morphometric parameters [9,10]. However, these methods rely heavily on large, manually annotated data sets [11–13]. Unsupervised learning methods have emerged to avoid costly manual labeling of images, some authors such as Caron et al., Zbontar et al., Chen et al., and Punn et al. [14–16] have used these tools in their research. These methods are based on coding, clustering, transfer learning, self-monitoring and other strategies.

The problem raised above generates great challenges to be solved. However, handling unlabeled images using unsupervised learning is not straightforward, and models trained with these techniques typically perform much less efficiently than supervised ones. But in recent years, unsupervised models have significantly narrowed the gap with supervised training, particularly with the recent achievements of contrastive and non-contrastive learning methods. Giving rise to self-supervised learning; which is a strategy that combines labeled data with unlabeled data during the training of a neural network. Initially, it learns unlabeled data features, then the weights are frozen and finally used in a tuning step to learn a specific task [17–20]. These architectures are based on generative approaches [21], predictive tasks [22], contrastive and non-contrastive learning [23] and bootstrap approaches [24,25].

In this paper, we develop a self-supervised artificial intelligence model to address the corneal endothelium segmentation problem. We use a large dataset of unlabeled images to train the encoder of a UNet network using the Barlow Twins approach to learn relevant data representations. Then with few labeled data, we train the UNet decoding path. In the following sections, we briefly review other related work, the proposed method, preliminary results, and conclusions.

2   Related Work

In recent years, due to the development of deep learning technologies, researchers have developed a great interest in computer-aided diagnostic systems to support healthcare services in different applications for classification, detection, and segmentation tasks [26]. This research will focus specifically on the task of segmentation of escular microscopy images of the corneal endothelium, due to the challenges present in this type of medical images when there is the presence of diseased cells.

Nowadays, many researches use models focused on supervised learning. Although it is a valuable technique in the segmentation of medical images of the corneal endothelium, it presents notable disadvantages compared to self-supervised and semi-supervised approaches. One of the main limitations of supervised learning is the large number of annotated images that the model requires for it to generalize. Not to mention that the acquisition of medical images is complex, they require expert personnel in the area and the labeled databases are limited. In contrast, self-supervised and semi-supervised approaches have the ability to memorize information from unlabeled data and learn a specific task with little annotation, thus addressing some of these limitations. Due to the above, these architectures become promising alternatives in image segmentation of the corneal endothelium.

In this context, Vigueras et al. [27] presented a fully automated method for estimating corneal endothelial parameters from specular microscopy images containing guttae. The proposed model was based on a DenseUNet neural network with nonlocal feedback attention to perform the semantic segmentation task. In general, the estimates agreed well with the reference values. The parameters were significantly better than those provided by commercial software, demonstrating the ability of this AI architecture to accurately estimate endothelial parameters even in the presence of edotheliopathies, like guttae in Fuchs' dystrophy.

Sierra et al. [3] proposed a UNet-based segmentation approach that requires minimal post-processing and achieves reliable CE morphometric assessment and guttae identification in all grades of Fuchs' dystrophy. They cast the segmentation problem as a regression task, using distance maps rather than a pixel-level classification task, as is typically done with the UNet architecture. These fully supervised approaches achieve decent performance but still require large annotated datasets.

There have been recent efforts in unsupervised or self-supervised methods have been developed to address the problem of the required annotation volume for data sets. In the area of biomedical image segmentation, self-supervised learning strategies can be grouped according to their approach as generative models [28], predictive tasks [22], contrastive learning [29], bootstrapping [24] and regularization [25].

Amodio et al. [30] presented the first fully unsupervised deep learning framework for medical image segmentation, which facilitated the use of the vast majority of image data that is not labeled or annotated. This unsupervised approach is based on a training objective with contrastive learning and self-coding aspects.

Previous contrastive learning approaches for medical image segmentation have focused on training at the image level. This approach is proposed at the patch level within the image (pixel-centric). This model achieves improved results in several critical medical imaging tasks, as verified by expert annotations on segmenting geographic atrophy regions from multi-subject retinal images.

Felfeliyan et al. [31] proposed an alternative self-supervised deep learning training strategy on unlabeled magnetic resonance imaging. In this research, they randomly applied different distortions to unlabeled image areas and then predicted the type of distortion and information loss. To do this, they used an improved version of the Mask-RCNN architecture to locate the location of the distortion and retrieve the pixels from the original image. This self-monitored pre-training improved the Dice Index score by 20% compared to training from scratch. The proposed self-supervised learning was simple, effective, and suitable for different ranges of medical image analysis tasks, including abnormality detection, segmentation and classification according to their complexity.

Therefore, the objective of this project is to investigate how unlabeled data can be used to pre-train a network and learn important data representations, then perform fine tuning for the segmentation task with few annotated corneal endothelial miscroscopy images. Unsupervised, semi-supervised, and self-supervised learning are becoming effective substitutes for transfer learning from large data sets.

3 Methods

3.1 Dataset Description

We utilized a dataset comprising 1300 in vivo images of corneal endothelial cells obtained from individuals with both healthy and pathological corneas. These images were captured at a resolution of 224×448 pixels. Among these images, we selected 230 patches measuring 96×96 pixels, which were annotated by domain experts. Additionally, we included 1719 patches of the same dimensions that lacked annotations. The acquisition of these images was performed using a Topcon SP3000P specular microscope equipped with Cell Count software. It's worth noting that the image collection process involved capturing images from either both eyes or just one eye, depending on the case. The study received approval from the ethics committee at Universidad Tecnologica de Bolivar, Colombia. Furthermore, due to the retrospective nature of the study, the requirement for informed consent was waived, in accordance with the principles outlined in the Declaration of Helsinki.

3.2 UNet Model

It is a convolutional neural network architecture that was designed for medical image segmentation. This model was originally developed Ronnenberger et al., in 2015 [32]. This architecture consists of two tracks. The first is that of contraction, also called encoder. It is used to capture the context of an image. The second way is symmetric expansion, called decoder. It also allows precise localization through transposed convolution. The mathematical model of the U-Net architecture is described by the following mathematical expressions:

$$w(\mathbf{x}) = w_c(\mathbf{x}) + w_0 \cdot \exp\left(-\frac{(d_1(\mathbf{x})) + d_2(\mathbf{x}))^2}{2\sigma^2}\right), \tag{1}$$

$$E = \sum_{x \in \Omega} w(\mathbf{x}) \cdot \log(P_{l(\mathbf{x})}(\mathbf{x})), \tag{2}$$

where $P_{l(\mathbf{x})}$ is the output of the softmax function, $d_1(\mathbf{x})$ and $d_2(\mathbf{x})$ indicate distances to the nearest boundary points, w_c represents the weight maps, w_0 and σ are constants.

3.3 Barlow Twins

It is a self-supervised learning method that applies redundancy reduction. It's to learn representations that are invariant to image distortions. It does not require large batches, gradient stops, momentum encoders and predictor networks. To overcome this problem, the Barlow Twins approach was proposed to pre-train the encoder in an unsupervised way, to then perform fine tuning, taking the weights of the pre-trained network and using them for the segmentation task with a limited number of samples annotated [33]. The mathematical model that describes the Barlow Twins method is as follows:

$$LBT = \sum_i (1 - C_{ij})^2 + \lambda \sum_i \sum_{j \neq i} C_{ij}^2, \tag{3}$$

$$C_{ij} = \frac{\sum_b * Z_{bi}^A * Z_{bj}^B}{\sqrt{\sum_b *(Z_{bi}^A)^2} * \sqrt{\sum_b *(Z_{bj}^B)^2}}, \tag{4}$$

where $\sum_i (1 - C_{ij})^2$ is an invariance term (diagonal or identity term) to drive neurons to produce the same output under different magnifications, and $\lambda \sum_i \sum_{j \neq i} C_{ij}^2$ is a redundancy reduction term (off-diagonal term) making each neuron produce a different output. The term λ is used to balance the contribution of the redundancy and invariance reduction terms.

3.4 Self-supervised Learning

Self-supervised learning is a machine learning paradigm that allows unlabeled data to be processed in order to obtain useful representations that can assist in subsequent learning tasks.

Due to the major challenges of medical databases, this strategy can help overcome the limitations of the availability of labeled data, allowing artificial intelligence models to capture important features and patterns more effectively.

This may lead to an improvement in the accuracy of medical image processing tasks, such as the segmentation task, which in turn may be beneficial for the early detection and diagnosis of ophthalmological diseases.

For this reason, this research proposes the implementation of an architecture based on self-supervised learning focused on the task of segmentation in images of the corneal endothelium, with the aim of improving the performance of CNN networks when there are no labels. In Fig. 1 we can see the block diagram of the implemented model.

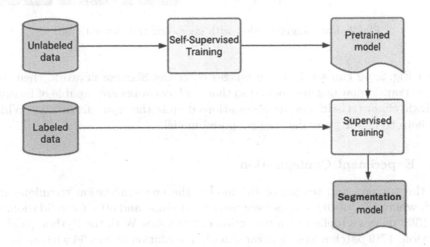

Fig. 1. Block diagram of the proposed SSL models.

In Fig. 1 we can detail the advantages that weakly supervised learning brings, because you can learn from many unlabeled images and do fine tuning with few annotated images.

3.5   Data Augmentation

The proposed model uses the geometric transformations strategy to mitigate the drawback of the absence and imbalance of annotations in the corneal endothelium image databases and is also part of the Siamese network strategy so that the model is more resistant to variations in the input data and generalize better. Particularly, this architecture applies a distortion to the input images randomly before the training stage, using the horizontal flip, vertical flip and rotations operations. In the following figure we can see the details.

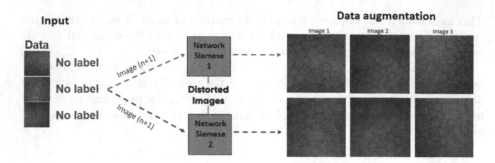

Fig. 2. Data augmentation with geometric transformations.

In Fig. 2, we can see how the images enter the Siamese network, then geometric transformations are applied so that both networks are capable of learning intrinsic characteristics and representations despite the input distortions. Which will help to better generalize the proposed model.

3.6   Experiment Configuration

For the training and testing of the model, the cross-validation technique was used, where 70% of the images were used for training and 30% for validation. Of the 1300 images supplied with resolution of 224×448, With the Python patchify function, 1719 patches were generated with a resolution of 96×96 without labels and 230 patches with the same resolution annotated by expert personnel. In order to increase and strengthen the feature maps of the unsupervised learning stage, without being required to annotate a very large volume of data, a higher proportion of unlabeled patches ($1719 \approx 88\%$) is provided. On the other hand, we worked with the Adam optimizer with a learning rate initialized at 1×10^{-3} for all the experiments, which had a decay of a factor of 0.1 once the learning stagnated to obtain better results of segmentation.

4   Proposed Approach

Despite the good results of CNN-based approaches, they generally exhibit limitations for modeling an explicit long-range relationship, due to the intrinsic locality of the convolution operations. Therefore, these architectures generally produce weak performance, especially for target structures that show large variations between images in terms of texture, shape, and size. To overcome this limitation, it is proposed to establish attention modules, blocks which have skip connections and transfer learning in the encoder of the proposed architecture. Which would allow us to model global and specific contexts. In the following figure we can see the architecture of our proposed model.

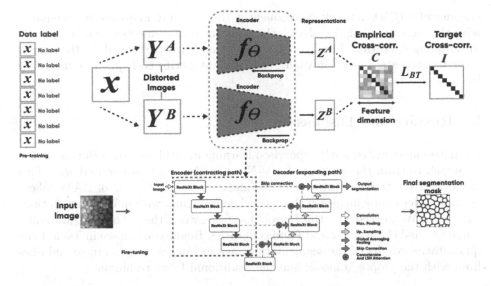

Fig. 3. Architecture of the proposed models for self-supervised learning training of the segmenter.

We aim to improve the predictions of the traditional UNet model. The proposed model addresses the challenge of limited data availability in two phases, as shown in Fig. 3. In phase one, a pre-training of the UNet encoder is performed with the Barlow Twins method, using the redundancy reduction principle to learn feature representations in an unsupervised way (without data annotations). Finally, for phase 2, fine-tuning is performed by taking the weights of the pretrained network and then using them for the semantic segmentation task with a limited number of annotated samples.

This proposal combines the advantages of the ResNet (Residual Network) which contains skip connections and subsequent residual blocks to extract semantic features that reduce the number of model parameters and improve the inference speed. Moreover, the benefits of the vision transformer (attention module) of the decoder, manage to combine multi-level functions and capture the global context.

We can see in Fig. 3 that the model receives a set of unlabeled images which enter a Siamese network. Then, a transformation is applied (rotations, translations, color changes, among others) and an encoder is introduced to generate the most representative feature maps using an empirical cross-correlation function. In the encoder stage, five filters [16, 32, 64, 128, 256], a global average pooling block, a fully connected layer block, ReLU activation, and normalization (FC+ReLU+BN) were used. These representations were frozen and became the encoder of a UNet network. Subsequently, these feature maps are concatenated with the corresponding decoder block, using the skip connections for the feature upsampling operation. Finally, a 1×1 convolution is performed on the output layer to generate a segmentation mask and categorize each pixel of the input images.

5   Results and Discussion

We have implemented a self-supervised learning model based on a Barlow Twins approach to train the encoder of a U-Net model in an unsupervised way. This model was pre-trained with a 32-core Linux server and 94 GB of RAM, where specular microscopy images of the corneal endothelium with different resolutions were used. We worked with the TensorFlow framework, the Python programming language, and the NumPy and Pandas libraries. Below you can see in Table 1 the quantitative results of the segmentation task in images of the corneal endothelium with the proposed model and the traditional UNet technique.

Table 1. Performance analysis of the UNet model and the proposed one [Accuracy (Acc), Precision (Pr), Dice Coefficient (DC) and mean Intersection-over-Union (mIoU)].

DS	Model	70% Training–30% Testing			
		Acc	Pr	DC	mIoU
1	UNet	0.8083	0.8420	0.8212	0.2797
	Ours	**0.8619**	**0.9393**	**0.9082**	**0.2939**

In Table 1 we can see that our proposed model presents better performance in the four metrics evaluated with respect to the UNet model. This can be corroborated in the segmentations obtained in Fig. 4. The mask predicted by our proposed method is quite similar to the reference mask, except in cases where images present problems of non-uniform illumination (limitations that will be covered in future approaches) and poor sharpness. These preliminary results demonstrate the benefit of using the pre-training strategy to improve the encoding stage, especially in cases where the availability to collect annotations is limited. In the following figure you can see the results obtained in the segmentation task in images of the corneal endothelium with healthy and diseased cells.

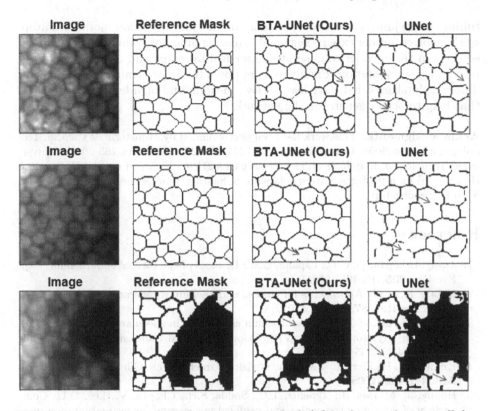

Fig. 4. Segmentation results. The proposed method defines better the intercellular boundary of the segmentation, despite some problems due to non-uniform illuminated areas.

In Fig. 4, you can see the qualitative performance of the self-supervised model compared to the traditional UNet model. Where it is evident that the proposed model improves its performance by freezing the weights of unsupervised training with unlabeled images, and then performing fine tuning with few images of the corneal endothelium with healthy cells and with Fuchs dystrophy. The images used present great challenges, due to their variations in scale, lighting, shadows, brightness, poor sharpness, among other aspects that make the training more complex.

6   Conclusion

We have developed a self-supervised learning model for the semantic segmentation task of images of the corneal endothelium obtained by specular microscopy to address one of the main challenges of the limited availability of annotated data. The results showed that the proposed strategy improves the segmentation performance of the UNet model. This improvement is evidenced due to encoder tuning, using jump connections, residual blocks, and attention modules. Future work involves further experiments and exploring different training strategies and settings in the encoder to generate better feature maps and ensure more accurate image segmentation. Finally, it is proposed to use pre-trained encoders with different databases that have learned low, medium and high level characteristics that help to better generalize the network.

Acknowledgments. This work has been partly funded by Ministerio de Ciencia, Tecnología e Innovacioón, Colombia, Project 124489786239 (Contract 763-2021), Universidad Tecnológica de Bolívar (UTB) Project CI2021P02. K. Mendoza and F. Quintero thank UTB for a post-graduate scholarship. S. Sanchez thanks Fondo Bicentenario for a Ph.D. Scholarship.

References

1. Jeang, L.J., Margo, C.E., Espana, E.M.: Diseases of the corneal endothelium. Exp. Eye Res. **205**, 108495 (2021)
2. Catala, P., et al.: Approaches for corneal endothelium regenerative medicine. Prog. Retin. Eye Res. **87**, 100987 (2022)
3. Sierra, J.S., et al.: Corneal endothelium assessment in specular microscopy images with Fuchs' dystrophy via deep regression of signed distance maps. Biomed. Opt. Express **14**(1), 335–351 (2023)
4. Knauer, C., Pfeiffer, N.: The value of vision. Graefes Arch. Clin. Exp. Ophthalmol. **246**, 477–482 (2008)
5. Huang, J., Maram, J., Tepelus, T.C., Sadda, S.R., Chopra, V., Lee, O.L.: Comparison of noncontact specular and confocal microscopy for evaluation of corneal endothelium. Eye Contact Lens Sci. Clin. Pract. **44**, 144–150 (2017)
6. Price, M.O., Fairchild, K.M., Price, F.W.: Comparison of manual and automated endothelial cell density analysis in normal eyes and DSEK eyes. Cornea **32**(5), 567–573 (2013). https://doi.org/10.1097/ico.0b013e31825de8fa
7. Luft, N., Hirnschall, N., Schuschitz, S., Draschl, P., Findl, O.: Comparison of 4 specular microscopes in healthy eyes and eyes with cornea guttata or corneal grafts. Cornea **34**(4), 381–386 (2015). https://doi.org/10.1097/ico.0000000000000385
8. Gasser, L., Reinhard, T., Böhringer, D.: Comparison of corneal endothelial cell measurements by two non-contact specular microscopes. BMC Ophthalmol. **15**, 87 (2015). https://doi.org/10.1186/s12886-015-0068-1
9. Piórkowski, A., Gronkowska–Serafin, J.: Towards precise segmentation of corneal endothelial cells. In: Ortuño, F., Rojas, I. (eds.) IWBBIO 2015. LNCS, vol. 9043, pp. 240–249. Springer, Cham (2015). https://doi.org/10.1007/978-3-319-16483-0_25
10. Selig, B., Vermeer, K., Rieger, B., Hillenaar, T., Luengo Hendriks, C.: Fully automatic evaluation of the corneal endothelium from in vivo confocal microscopy. BMC Med. Imaging **15**, 13 (2015). https://doi.org/10.1186/s12880-015-0054-3
11. Shilpashree, P., Kaggere, S., Sudhir, R., Srinivas, S.: Automated image segmentation of the corneal endothelium in patients with Fuchs dystrophy. Transl. Vis. Sci. Technol. **10**, 27 (2021). https://doi.org/10.1167/tvst.10.13.27
12. Daniel, M., et al.: Automated segmentation of the corneal endothelium in a large set of 'real-world' specular microscopy images using the U-net architecture. Sci. Rep. **9** (2019). https://doi.org/10.1038/s41598-019-41034-2

13. Vigueras-Guillén, J., et al.: DenseUNets with feedback non-local attention for the segmentation of specular microscopy images of the corneal endothelium with guttae. Sci. Rep. **12** (2022). https://doi.org/10.1038/s41598-022-18180-1
14. Caron, M., et al.: Emerging Properties in Self-supervised Vision Transformers (2021)
15. Zbontar, J., Jing, L., Misra, I., LeCun, Y., Deny, S.: Barlow twins: self-supervised learning via redundancy reduction. In: International Conference on Machine Learning, pp. 12310–12320. PMLR (2021)
16. Punn, N.S., Agarwal, S.: BT-Unet: a self-supervised learning framework for biomedical image segmentation using Barlow twins with U-net models. Mach. Learn. **111**(12), 4585–4600 (2022)
17. Jiao, R., Zhang, Y., Ding, L., Cai, R., Zhang, J.: Learning with Limited Annotations: A Survey on Deep Semi-supervised Learning for Medical Image Segmentation (2022)
18. Wu, Y., et al.: Mutual Consistency Learning for Semi-supervised Medical Image Segmentation (2022)
19. Fang, K., Li, W.-J.: DMNet: difference minimization network for semi-supervised segmentation in medical images. In: Martel, A.L., et al. (eds.) MICCAI 2020. LNCS, vol. 12261, pp. 532–541. Springer, Cham (2020). https://doi.org/10.1007/978-3-030-59710-8_52
20. Chen, S., Bortsova, G., Juarez, A.G.-U., Tulder, G., Bruijne, M.: Multi-task Attention-Based Semi-supervised Learning for Medical Image Segmentation (2019)
21. Devlin, J., Chang, M.-W., Lee, K., Toutanova, K.: BERT: pre-training of deep bidirectional transformers for language understanding. arXiv preprint arXiv:1810.04805 (2018)
22. Pathak, D., Krähenbühl, P., Donahue, J., Darrell, T., Efros, A.A.: Context encoders: feature learning by inpainting. In: 2016 IEEE Conference on Computer Vision and Pattern Recognition (CVPR), pp. 2536–2544 (2016). https://doi.org/10.1109/CVPR.2016.278
23. Balestriero, R., LeCun, Y.: Contrastive and Non-contrastive Self-supervised Learning Recover Global and Local Spectral Embedding Methods (2022)
24. Grill, J.-B., et al.: Bootstrap your own latent: a new approach to self-supervised learning (2020)
25. Ghosh, S., Seth, A., Mittal, D., Singh, M., Umesh, S.: DeLoRes: Decorrelating Latent Spaces for Low-Resource Audio Representation Learning (2022)
26. Wang, R., Lei, T., Cui, R., Zhang, B., Meng, H., Nandi, A.K.: Medical image segmentation using deep learning: a survey. IET Image Proc. **16**(5), 1243–1267 (2022). https://doi.org/10.1049/ipr2.12419
27. Vigueras-Guillén, J.P., et al.: Fully convolutional architecture vs sliding-window CNN for corneal endothelium cell segmentation. BMC Biomed. Eng. **1**, 4 (2019). https://doi.org/10.1186/s42490-019-0003-2
28. Devlin, J., Chang, M.-W., Lee, K., Toutanova, K.: BERT: Pre-training of Deep Bidirectional Transformers for Language Understanding (2019)
29. Noroozi, M., Favaro, P.: Unsupervised Learning of Visual Representations by Solving Jigsaw Puzzles (2017)
30. Liu, C., et al.: CUTS: A Fully Unsupervised Framework for Medical Image Segmentation (2023)
31. Felfeliyan, B., et al.: Self-supervised-RCNN for Medical Image Segmentation with Limited Data Annotation (2022)

32. Ronneberger, O., Fischer, P., Brox, T.: U-net: convolutional networks for biomedical image segmentation. In: Navab, N., Hornegger, J., Wells, W.M., Frangi, A.F. (eds.) MICCAI 2015. LNCS, vol. 9351, pp. 234–241. Springer, Cham (2015). https://doi.org/10.1007/978-3-319-24574-4_28
33. Marsocci, V., Scardapane, S.: Continual Barlow twins: continual self-supervised learning for remote sensing semantic segmentation. IEEE J. Sel. Top. Appl. Earth Obs. Remote Sens. **16**, 5049–5060 (2023). https://doi.org/10.1109/JSTARS.2023.3280029

Tuberculosis Drug Discovery Estimation Process by Using Machine and Deep Learning Models

Michael S. Ramirez Campos[1,2] (iD), Diana C. Rodríguez[2] (iD),
and Alvaro D. Orjuela-Cañón[2(✉)] (iD)

[1] Universidad Escuela Colombiana de Ingeniería Julio Garavito, Bogota, D.C., Colombia
[2] School of Medicine and Health Sciences, Universidad del Rosario, Bogota, D.C., Colombia
`alvaro.orjuela@urosario.edu.co`

Abstract. Tuberculosis is a contagious disease considered as world emergency by the World Health Organization. One of the common prevalent problems are associated to drug-resistant TB, because of unsuccessful treatments of using antibiotics. The use of artificial intelligence algorithms, mainly machine learning (ML) models have allowed to provided more tools for the drug discovery field. For this study, the methodology used was driven to identify new components that may contribute to the inhibition of the *inhA* protein. Leveraging ML models that learn from data, six regression models were implemented. Best model obtained R2 value of 0.99 and a MSE value of 1.8 e-5.

Keywords: Tuberculosis · Drug-Resistant · Machine Learning · Drug-Discovery

1 Introduction

Tuberculosis (TB) is a contagious disease considered as world emergency by the World Health Organization (WHO). Before the Covid-19 pandemic, TB was the first cause of death above the AIDS. Annually, there are more than one and a half million of deaths caused by TB [1]. In spite to different efforts around the globe, currently strategies as Stop TB from WHO are more developed to eradicate this disease [1, 2]. One of the common prevalent problems are associated to drug-resistant TB, because of unsuccessful treatments of using antibiotics. Issues become worse for the multidrug-resistant TB is present, where more medicaments and lines of treatment are not enough to treat the disease. This population is growing in at least a 6% in last years [3, 4].

Antibiotics are drugs whose main purpose is to control the infections by bacteria. There, the infection is inhibited by metabolic processes, where some of these mechanisms exterminate the bacteria, for example, destroying the cell wall. In this case, the antibiotic can affect the synthetize process of the structural proteins, or preventing the necessary transportation of precursors, mainly. However, these drugs can affect the cytoplasm, enzyme proteins, and nucleic acid synthesis [5]. Despite of effectiveness of these substances, it is important to follow the recommendations in terms of quantities times provided for the healthcare professionals. In case of treatment is not complete or not seriously considered, the bacteria can develop resistance to the drug [4].

© The Author(s), under exclusive license to Springer Nature Switzerland AG 2024
A. D. Orjuela-Cañón et al. (Eds.): ColCACI 2023, CCIS 1865, pp. 43–53, 2024.
https://doi.org/10.1007/978-3-031-48415-5_4

For TB, many years of researching have provided useful information beyond the clinical aspects. From the genomic data, genes and its expression have provided important information. For example, the *inhA* gene codifies for enoyl acyl reductase, an enzyme protein involved in the synthesis of fatty acids, mainly in the biosynthesis of mycolic acid. This protein is a promising target for anti-TB drug development because the conserved active sites, which have not been identified in other TB targets [6]. It is the main target of the anti-TB drug isoniazid, where different mutations of the gene can cause resistance to isoniazid [7, 8]. For this study, the methodology used was driven to identify new components that may contribute to the inhibition of the *inhA* protein.

The use of artificial intelligence algorithms, mainly machine learning (ML) models have allowed to provided more tools for the drug discovery field [9]. Furthermore, information extracted from chemical compounds datasets can be useful for training ML models and obtain new proposals for drug design, at applying algorithms with regression tasks [10].

Examples of drug discovery are utilized in different fields. In [11], a prediction of the minimum inhibition concentration (MIC) value reported in the literature for 24 drug molecules used to treat tuberculosis, using multiple regression methods with a R^2 of 0.74 as the best result. There, the quantitative structure-analysis relationship (QSAR) method was employed. The QSAR analysis considers the chemical structure of different compounds and computes a quantitative correlation to a biology activity, showing properties of inhibition of the cell [12]. Similarly, in [13] a 3D-QSAR study was carried out for pyrrole antifungal derivatives with the objective of relating chemical structures with inhibitory activity in TB. The model developed in this study presented an R^2 value of 0.86 when QSAR features were combined with those obtained by comparative molecular field analysis (coMFA).

On the other hand, molecular descriptors calculated from paDEL-descriptors were employed in order to propose a QSAR model that allows a rapid prediction of the inhibitory activity against the 3CLpro enzyme [14, 15]. The model was generated using the partial least squares regression method, and obtained an R^2 value of 0.756 as well as a mean absolute error (MAE) of 0.374. A comparison of different regression methods, such as random forest, stochastic gradient boosting, multivariate adaptive regression splines, and Gaussian processes regression, was performed to generate predictors of the bioactivity of different heterocyclic compounds. In this study, the best model was obtained by using a random forest, where an R^2 value of 0.862 was reported [16]. Finally, the same team of this proposal worked in the study of regression models for the inhibition of staphylococcus aureus quinolone protein were analyzed, with a random forest model as the best one, with a R^2 value of 0.866 [17].

Inside the machine learning field, the deep learning (DL) treats the specific neural networks with bigger architectures than traditional ones. In relation to the drug discovery area, some works have exhibited the advantages in detailed areas such as prediction of bioactivity or Mode-of-Action and analysis, and primary and secondary drug screening as topics associated to the present work [18]. In addition, DL models have been employed to predict the drug-target interaction and de novo drug design based on benchmark datasets and tools for modeling and training [19]. For problems associated to TB, DL models are in the future of the drug discovery in this disease and how the applications include molecular docking, molecular dynamics simulations, and computational hit analysis, employing *in-silico* proposals [20, 21].

The present proposal disserts about the use machine and deep learning models for bioactivity prediction based on different available datasets. Compounds related to the *inhA* expressed gene were considered for the analysis, searching to contribute in the drug-resistant TB problem.

2   Materials and Methods

For the molecular compounds election, it is important to determine a dataset with information about the molecules employed in the pharmaceutical industry. There are some of them that are available to performance computational experimentation. In this study, we followed this issue, by employing data without restrictions. Then, the computational experimentation is described with the procedure to compute the features and the machine and deep learning used models.

2.1 Dataset

The ChEMBL database provides information related to chemistry, bioactivity and genomic properties of bioactive molecules. There, data from studies found in peer-reviewed literature in medicinal chemistry journals is considered and clustered in available information. This database allows to know several specific properties, which were obtained from the characterization of compounds used for the manufacture of drugs, and which can be used in the production of new molecules through the drug discovery [22].

Since this study will work based on three bioactivity units, 1104 compounds were identified, where 569 were reported in percentage inhibition values, 413 in MIC units and 122 in 50% inhibition concentration units (IC50), using the enoyl acyl – reductase protein (encoded by the *inhA* gene) as a target.

2.2 Features Computation

In order to obtain data to be used in the training of the ML models, a choice of MIC and IC50 values was developed. For this purpose, the information provided in the literature was taken into account, where for both cases three classifications are generated for the molecule, according to the concentration needed to be effective. In this sense, the three classifications allowed establishing three labels for the data: sensitive, indeterminate (intermediate), and resistant [23, 24]. In addition, it is important to note that a standardization was applied by using the logarithmic scale for MIC and IC50 values, improving the range of values. The output value used in the ML regression models was $-log(MIC)$ and $-log(IC50)$.

Crystopher A. Lipinski formulated the Lipinski rules in 1997, based on empirical observation on the behavior of oral drugs. These five rules represent the parameters that a drug must meet to determine its oral viability. Aspects taken into account of these rules are: molecular mass less than 500 Daltons, high lipophilicity (expressed as LogP less than 5), less than 5 hydrogen bond donors, less than 10 hydrogen bond acceptors, and molar refractivity between 40–130 [23]. In the present case, the molecular weight, octanol-water partition coefficient (logp), hydrogen bond donors and acceptors (4 out the 5 Lipinski parameters) were calculated from each compound. Since there are only four features for each molecule, a feature reduction step was not considered.

2.3 Machine Learning Models

Leveraging ML models that learn from data, five regression models were implemented based on the scientific kit library of Python [25]. The considered models were: a multi-layer perceptron (MLP), support vector repressors (SVR), gradient boosting regressors (GBR), random forest (RF), and k-nearest neighbors (kNN). The models were chosen according the high level of employability in previous results in similar studies. Each one of them was analyzed for the three data sets, according to the different bioactivity units: *i)* set of MIC units, *ii)* set of inhibition percentage, and *iii)* set of IC50 units.

The hyperparameters for training of the ML models were obtained by using a heuristical method and a five-fold cross-validation technique. A portion of 80% of the data was used for training and the remaining set was used to test the models. To calculate the error, the R^2 value was taken into account. In addition, the mean square error (MSE) and mean average error (MAE) were examined, according to the possibility of finding differences in the results given the metric to compute de error.

2.4 Deep Learning Models

Inside the DL applications associated to analysis of compounds and provide information for drug discovery, a wide range of models can be employed. However, based on previous reported works in this field, EfficientNet is a model used for the mechanism of action associated to molecular data from Broad Institute [26, 27]. Based on other DL model, the MobileNet architecture was employed to identify drugs through images. There, authors used a Faster convolutional neural network and a single shot detector based on MobileNet and the OpenVINO library [28, 29]. The ResNet architecture was employed for the aqueous solubility prediction on novel compounds for solubility predictions, compared to shallow networks of three to seven layers [30, 31].

According to the described applications, the present work used these three architectures to evaluate the estimators, in a similar way than the basic ML. In addition, measures of R^2, MSE, and MAE were employed to analyze the models.

3   Results

Tables 1 and 2 resume the results obtained for each ML and DL model, according to the used units (MIC, IC50 and inhibition). Each table shows the performance of each regressor for the three scenarios, according to the average value of R^2, MSE and MAE, and its standard deviation. Figures 1, 2 and 3 visualize the regression values for the three considered scenarios.

Table 1. Results for the regressors based on ML models.

Scenario	Regressor	R2 Score	MAE Score	MSE Score
MIC Units Set	SVR	0.999	7.56e-03	4.69e-04
	MLP	0.999	3.06e-02	7.78e-03
	GBR	0.997	2.63e-02	7.69e-03
	Random Forest	0.995	5.54e-02	1.52e-02
	KNN	0.486	0.776	1.402
IC50 Units Set	SVR	0.999	7.34e-03	9.01e-05
	MLP	0.999	1.85e-02	3.99e-03
	GBR	0.999	1.33e-02	1.16e-03
	Random Forest	0.997	2.98e-02	7.63e-03
	KNN	0.561	0.805	1.473
Inhibition Percentage Unit Set	SVR	0.999	4.81e-03	3.16e-05
	MLP	0.999	0.118	6.35e-02
	GBR	0.999	8.22e-02	5.24e-02
	Random Forest	0.999	0.235	0.126
	KNN	0.966	3.311	27.035

Table 2. Results for the regressors based on DL models.

Scenario	Regressor	R2 Score	MAE Score	MSE Score
MIC Units Set	ResNet	0.99	**3.23e-02**	**1.72e-03**
	MobileNet	0.93	0.12	0.17
	EfficientNet	0.99	8.95e-02	2.89e-02
IC50 Units Set	ResNet	0.95	0.12	0.12
	MobileNet	0.96	**9.56e-02**	**0.1**
	EfficientNet	0.94	0.11	0.16
Inhibition Percentage Unit Set	ResNet	0.95	**2.44**	**24.11**
	MobileNet	0.95	3.12	24.4
	EfficientNet	0.92	3.23	44.73

4 Discussion

Initially, it is possible to determine that the expected results were achieved. This, because the tables and in figures represent the performance of the vast majority of the generated models. In addition, the different scenarios of datasets obtained metrics close to the ideal values (diagonal performance, see Figs. 1, 2 and 3), and the R^2 values above 0.99. In

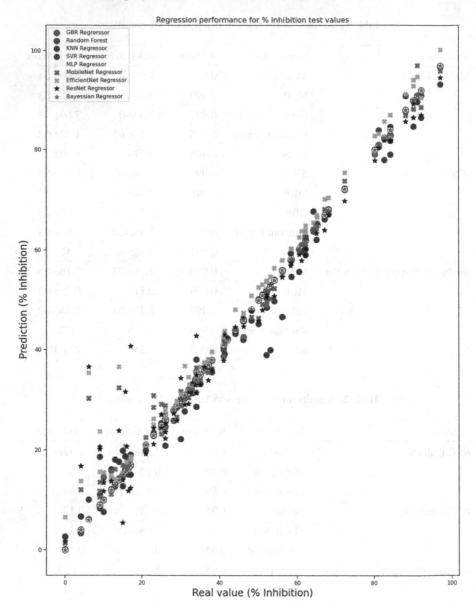

Fig. 1. Regression results for the molecules with inhibition percentage unit set

terms of MAE and MSE, its values were close to zero. It should be remembered that the MAE and the MSE are global measures related to the error presented by each model when making predictions, and R^2 value quantifies the fit of the predictions made by the models with respect to the real values. MAE and MSE values should be equal to zero, and R^2 value should be close to one.

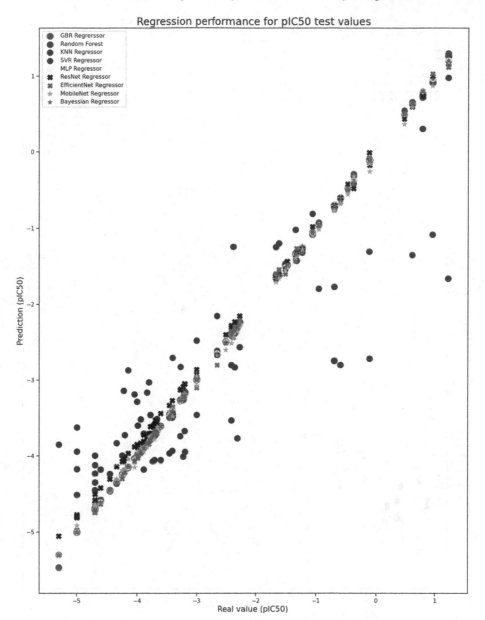

Fig. 2. Regression results for the molecules with IC50 unit set

Likewise, the results obtained in this study exceed the results of other reported works. The reported literature exhibits how the use of DL models can improve the results, for example in solubility prediction [30]. There, for the regression, a ResNet neural network with 26 layers was better than shallow neural network, getting better results with a 20% more. Directly related to TB, based on molecular data, the accuracy made progress from

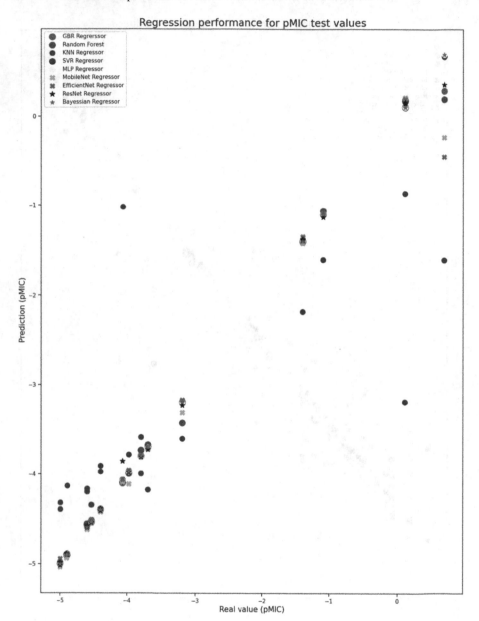

Fig. 3. Regression results for the molecules with MIC unit set

62% at using an MLP to 81%, employing an EfficientNet [26]. This reflects that more precise models were created, and demonstrates that the methodology carried out carried out allows the development of prediction models for different units of bioactivity in a reliable way.

An important limitation has to be noted here, according to the size of the used datasets because it could be smaller than reported with tens of thousands elements. This was reported as a pitfall of the use of ML for accelerate the drug discovery processes [20]. The number of considered compounds must be higher, planning future work in this aspect.

Finally, it is important to clarify that the Lipinski's descriptors allow us to know the structural properties of the molecules. For this reason, they were considered as input in the ML models. Furthermore, these characteristics proved to be functional to carry out the development of bioactivity prediction models for drug-resistant tuberculosis. Most of the used models yielded promising results with only four of the five descriptors as input, which allowed to consider them as relevant information for future work related to the topic addressed in this study. In addition, the PaDEL descriptors could be explored in future experiments, according to the advantages of the bigger neural networks models.

5 Conclusions

The proposed methodology has demonstrated its effectiveness in developing prediction models for different units of bioactivity of molecules, which can potentially address drug-resistance of these mycobacteria. In this sense, the results presented in this study are promising, since, as mentioned above, the values of the performance metrics of the models are close to the ideal values, and likewise, they exceed the works reported in the state of the art.

In addition, it was found that Lipinski descriptors could be useful as valuable features when generating these models. This constitutes a valuable contribution for future work, taking into account that, with four molecular descriptors, the computational cost related to the use of other types of characteristics (e.g. fingerprints) is considerably reduced.

Finally, the proposed method is considered innovative, as there is limited literature on the application of machine learning techniques for predicting bioactivity values associated to TB and the drug-resistant challenge. However, future work should focus on acquiring a larger dataset to develop models that are more robust.

Acknowledgement. Authors acknowledge the support of the Universidad del Rosario for funding this project. In addition, the contribution of research incubator team *Semillero en Inteligencia Artificial en Salud: Semill-IAS* and incubator *SyNERGIA*.

References

1. Harding, E.: WHO global progress report on tuberculosis elimination. Lancet Respir. Med. **8**, 19 (2020)
2. Chakaya, J., et al.: The WHO global tuberculosis 2021 report–not so good news and turning the tide back to end TB. Int. J. Infect. Dis. **124**, S26–S29 (2022)
3. Lange, C., Dheda, K., Chesov, D., Mandalakas, A.M., Udwadia, Z., Horsburgh, C.R.: Management of drug-resistant tuberculosis. Lancet**394**, 953–966 (2019)
4. Khawbung, J.L., Nath, D., Chakraborty, S.: Drug resistant tuberculosis: a review. Comp. Immunol. Microbiol. Infect. Dis.**74**, 101574 (2021)

5. Uddin, T.M., et al. Antibiotic resistance in microbes: history, mechanisms, therapeutic strategies and future prospects. J. Infect. Public Health **14**, 1750–1766 (2021)
6. Prasad, M.S., Bhole, R.P., Khedekar, P.B., Chikhale, R.: V Mycobacterium enoyl acyl carrier protein reductase (InhA): a key target for antitubercular drug discovery. Bioorg. Chem. **115**, 105242 (2021)
7. Wang, X., et al.: Intrabacterial metabolism obscures the successful prediction of an InhA inhibitor of mycobacterium tuberculosis. ACS Infect. Dis. **5**, 2148–2163 (2019)
8. Yao, C., et al.: Detection of rpoB, katG and inhA gene mutations in Mycobacterium tuberculosis clinical isolates from Chongqing as determined by microarray. Clin. Microbiol. Infect. **16**, 1639–1643 (2010)
9. Dara, S., Dhamercherla, S., Jadav, S.S., Babu, C.H.M., Ahsan, M.J.: Machine learning in drug discovery: a review. Artif. Intell. Rev. **55**, 1947–1999 (2022)
10. Vamathevan, J., et al.: Applications of machine learning in drug discovery and development. Nat. Rev. Drug Discov. **18**, 463–477 (2019)
11. Dwivedi, N., Mishra, B.N., Katoch, V.M.: 2D-QSAR model development and analysis on variant groups of anti-tuberculosis drugs. Bioinformation **7**, 82 (2011)
12. Cherkasov, A., et al.: QSAR modeling: where have you been? Where are you going to? J. Med. Chem. **57**, 4977–5010 (2014)
13. Ragno, R., et al.: Antimycobacterial pyrroles: synthesis, anti-mycobacterium tuberculosis activity and QSAR studies. Bioorganic Med. Chem. **8**, 1423–1432 (2000)
14. Yap, C.W.: PaDEL-descriptor: an open source software to calculate molecular descriptors and fingerprints. J. Comput. Chem. **32**, 1466–1474 (2011)
15. De, P., Bhayye, S., Kumar, V., Roy, K.: In silico modeling for quick prediction of inhibitory activity against 3CLpro enzyme in SARS CoV diseases. J. Biomol. Struct. Dyn. **40**, 1010–1036 (2022)
16. Kryshchyshyn, A., Devinyak, O., Kaminskyy, D., Grellier, P., Lesyk, R.: Development of predictive QSAR models of 4-thiazolidinones Antitrypanosomal activity using modern machine learning algorithms. Mol. Inform. **37**, 1700078 (2018)
17. Campos, M.S.R., López, D.A.G., Rivera, J.A.C., Rodriguez, D.C., Perdomo, O.J., Orjuela-Cañon, A.D.: Bioactivity predictors for the inhibition of staphylococcus aureus quinolone resistance protein. In: Proceedings of the Applied Computer Sciences in Engineering: 9th Workshop on Engineering Applications, WEA 2022, Bogotá, Colombia, November 30--December 2, 2022, Proceedings; pp. 31–40 (2022)
18. Gupta, R., Srivastava, D., Sahu, M., Tiwari, S., Ambasta, R.K., Kumar, P.: Artificial intelligence to deep learning: machine intelligence approach for drug discovery. Mol. Divers. **25**, 1315–1360 (2021)
19. Kim, J., Park, S., Min, D., Kim, W.: Comprehensive survey of recent drug discovery using deep learning. Int. J. Mol. Sci. **22**, 9983 (2021)
20. Winkler, D.A.: The impact of machine learning on future tuberculosis drug discovery. Expert Opin. Drug Discov. **17**, 925–927 (2022)
21. Kingdon, A.D.H., Alderwick, L.J.: Structure-based in silico approaches for drug discovery against Mycobacterium tuberculosis. Comput. Struct. Biotechnol. J. **19**, 3708–3719 (2021)
22. Mendez, D., et al.: Others ChEMBL: towards direct deposition of bioassay data. Nucleic Acids Res. **47**, D930–D940 (2019)
23. Benet, L.Z., Hosey, C.M., Ursu, O., Oprea, T.I.: BDDCS, the rule of 5 and drugability. Adv. Drug Deliv. Rev. **101**, 89–98 (2016)
24. Lima, A.N., Philot, E.A., Trossini, G.H.G., Scott, L.P.B., Maltarollo, V.G., Honorio, K.M.: Use of machine learning approaches for novel drug discovery. Expert Opin. Drug Discov. **11**, 225–239 (2016)
25. Pedregosa, F., et al.: Scikit-learn: machine learning in python. J. Mach. Learn. Res. **12**, 2825–2830 (2011)

26. Tian, G., Harrison, P.J., Sreenivasan, A.P., Carreras-Puigvert, J., Spjuth, O.: Combining molecular and cell painting image data for mechanism of action prediction. Artif. Intell. Life Sci. **3**, 100060 (2023)
27. Koonce, B., Koonce, B.: EfficientNet. convolutional neural networks with swift Tensorflow image recognit. Dataset Categ. 109–123 (2021)
28. Biswas, R., Basu, A., Nandy, A., Deb, A., Haque, K., Chanda, D.: Drug discovery and drug identification using AI. In: Proceedings of the 2020 Indo--Taiwan 2nd International Conference on Computing, Analytics and Networks (Indo-Taiwan ICAN), pp. 49–51 (2020)
29. Sinha, D.; El-Sharkawy, M. Thin mobilenet: an enhanced mobilenet architecture. In: Proceedings of the 2019 IEEE 10th Annual Ubiquitous Computing, Electronics & Mobile Communication Conference (UEMCON), pp. 280–285 (2019)
30. Cui, Q., et al.: Improved prediction of aqueous solubility of novel compounds by going deeper with deep learning. Front. Oncol. **10**, 121 (2020)
31. Wightman, R., Touvron, H., Jégou, H.: Resnet strikes back: an improved training procedure in timm. arXiv Prepr. arXiv2110.00476 (2021)

Biological Applications

A Robust Hybrid Control Approach Tuned by PSO for Long-Time Delay Nonlinear Chemical Processes

Marco Herrera, Diego S. Benítez, Noel Pérez-Pérez, Antonio Di Teodoro, and Oscar Camacho

Colegio de Ciencias e Ingenierías "El Politecnico", Universidad San Francisco de Quito USFQ, Quito 170157, Ecuador
marco.herrera@ieee.org, {dbenitez,nperez,nditeodoro,ocamacho}@usfq.edu.ec

Abstract. This paper proposes a hybrid controller that mixes the Smith Predictor scheme along the sliding mode method and uses particle swarm optimization (PSO) to fine-tune the controller parameters. The proposed approach considers including the Taylor series approximation to put together a dead-time term as the non-invertible part and a gain for the invertible parts; in this way, the controller synthesis is simple. The performance of the designed controller is measured using the error-dependent performance indexes using a nonlinear chemical reactor with long-time delay systems. The results show that the proposed controller tuned using PSO can significantly increase the efficiency of designing robust controllers for nonlinear systems.

1 Introduction

Time delays are prevalent within industrial processes, as highlighted by Ming et al. (2018) [15]. They manifest when mass or energy is transported through a medium, as discussed in the context of control systems by Normey [18]. Time-delay systems can be found in various domains, including transmission systems, chemical processes, metallurgical operations, hydraulic and pneumatic systems, power grids, biological systems, environmental processes, and ecosystems, as noted in [3]. The presence of time delays introduces intricacies in the analysis and control of these systems, necessitating specialized attention [24,27].

Process industries frequently employ the proportional integral derivative (PID) control algorithm due to its ease of use, robustness, and effective practical implementation [1,19]. However, the control performance achieved with a PID controller is constrained when the process has extended the dead time. Therefore, it has been discovered that the PID controller would become less effective due to a considerable increase in process dead time [22].

The prevailing control strategy used in long delay systems is the Smith predictor, as discussed in [14]. This approach integrates a PID controller with an inner

The Universidad San Francisco de Quito supported this work through the Poli-Grants Program under Grant 17965.

A. D. Orjuela-Cañón et al. (Eds.): ColCACI 2023, CCIS 1865, pp. 57–71, 2024.
https://doi.org/10.1007/978-3-031-48415-5_5

loop incorporating a process model to minimize the adverse effects of prolonged delay times within the system. Although the Smith predictor is a model-based controller renowned for its efficacy in managing processes with significant dead times, it exhibits sensitivity to modeling inaccuracies, as highlighted by Normey (2007) [17]. Theoretically, Smith predictor control presents a promising solution to the challenges of controlling time-delay systems. However, its practical performance enhancement in real-time applications is contingent on the robustness of the control system in the face of modeling errors [2] and recently explored by Espin et al. (2023) [6].

In the industrial world, disparities commonly emerge between the actual plant and the mathematical model crafted for controller design. These disparities stem from unaccounted for dynamics and fluctuations in system parameters. Consequently, engineers must prioritize the development of controllers capable of robust operation, withstanding parameter variations effectively, and ensuring their efficacy in real-world scenarios.

The Sliding-Mode Control technique is one of the robust nonlinear control methods widely utilized to control uncertain nonlinear systems. Despite its intrinsic capacity to deal with uncertainties, it suffers from being a model-based nonlinear method. One must have an appropriate system model to apply this strategy to nonlinear control. With the prowess of SMC in resolving model-plant discrepancies, controllers have been designed using this method [21,25].

Sliding mode control (SMC) is a robust strategy designed to regulate systems with model uncertainty and is immune to outside disturbances. However, chattering, a high-frequency oscillation that occurs in the output control action and is a weakness of the SMC, can damage actuators (mechanical parts), activate unmodeled high-frequency dynamics, and deteriorate the overall system, leading to sudden instability.

Considering the previous advantages of the Smith Predictor for long delay systems, the robustness of the Sliding Mode for uncertainties, and the particle swarm optimization algorithm, this work proposes a hybrid controller that mixes the Smith Predictor scheme along the Sliding Mode Methodology and uses Particle Swarm Optimization to the best tuning parameters for the controller. Hybrid control is a control alternative that allows controllers to be designed to work under larger and harder operating conditions than simple controllers. They are a combination of simple control approaches to obtain a new controller that takes advantage of the positive characteristics of each one, which would not be achieved if it were a single controller [6,21]; each of them in its range of operation is designed to handle a manipulated variable and thus contribute to the hybrid scheme. A previous version applied to linear systems can be found in [8].

The paper is organized as follows: Sect. 2 describes some fundamentals. Section 3 shows the controllers designed, in Sect. 4, the simulation results applied by simulation in a linear system with a long delay, and finally, the conclusions.

2   Preliminaries

2.1   First Order Plus Deadtime Model of Process

Due to the difficulties that most industrial processes present in mathematical analysis, the most common model used in these processes to design controllers is the first order plus dead time (FOPDT) model [22].

$$\frac{Y(s)}{U(s)} = \frac{K}{(\tau s + 1)} e^{-t_0 s} \tag{1}$$

$Y(s)$: Transmitter output; $U(s)$: Control action; K is static gain, t_0 is a time delay, and τ is a time constant. The approximation of the FOPDT model can represent many industrial processes [5].

To obtain the characteristic parameters of FOPDT, the process curve method suggested by Smith and Corripio [22] is used. Following the procedure, the parameters are calculated with the following expressions [13]:

$$K = \Delta Y / \Delta U \tag{2}$$

$$\tau = 1.5\,(t_{63.2\%} - t_{28.3\%}) \tag{3}$$

$$t_0 = t_{63.2\%} - \tau \tag{4}$$

2.2   Dead Time Compensator or Smith Predictor

The Smith predictor (SP), which contains a process model without delay in its structure, was the first control method and one of the most widely used in the industry to account for time delays [14]. However, the SP has revealed a few flaws related to the accuracy of the built model plant. As a result, the performance of closed-loop control methods is decreased in actual applications due to parameter uncertainties, internal/external disturbances, and delay fluctuations.

2.3   Sliding Mode Control

The sliding mode control is derived from the variable control structure (VCS) [25]. SMC is a robust control that responds well to nonlinearities, delays, uncertainties, disturbances, and modeling errors [5,6].

The procedure for designing this type of controller can be found in [5,25]. The sliding mode controller contains two parts: a discontinuous part $u_d(t)$ responsible for the reachability mode and the equivalent or continuous part $u_{eq}(t)$ associated with the sliding mode. The control law that governs the SMC is as follows.

$$u_{SMC}(t) = u_{eq}(t) + u_d(t). \tag{5}$$

The discontinuous part, $u_d(t)$, is nonlinear and includes a control law switching element:

$$u_d(t) = K_D \frac{\sigma(t)}{|\sigma(t)| + \delta} \tag{6}$$

where K_D is a tuning parameter related to the reachability phase and δ is used to reduce the chattering problem [5].

3   The Hybrid Control Approach Synthesis

In this section, a hybrid control is designed based on the sliding-mode control and the Smith predictor approach. The parameters of the proposed controller are tuned using the particle swarm optimization (PSO) method; see Fig. 1.

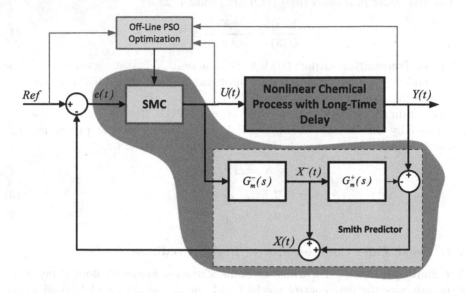

Fig. 1. Hybrid Control with Smith predictor scheme and PSO optimization.

To design and adjust a control system, it is necessary to know how the dynamic behavior of the process is regulated. Therefore, a mathematical process model must be developed to have this information. Additionally, this model must provide accurate information to estimate the effects of a control system on the behavior of a controlled process at the operating point. More specifically, this model needs to forecast the impact on the output variable of the control process for servo- and regulatory control [7].

Consider $f(x) = e^x$ and an arbitrarily positive real number R such that $0 \leq f^n(x) \leq e^R$, where $n \geq 0$ for all $x \in (-R, R)$. In particular, for $c = 0$ we get the Taylor expansion for the exponential $e^x = \sum_{n=0}^{\infty} \frac{1}{n!} x^n$.

Considering the Taylor approximation of the first order, we have the following.

$$\frac{1}{\tau s + 1} \cong e^{-\tau s} \tag{7}$$

According to the structure of the Smith predictor shown in Fig. 1, thus (8) can be separated into invertible and noninvertible parts, such as:

$$G_m(s) = G_m^-(s) G_m^+(s). \tag{8}$$

where:

$$G_m^-(s) = K, \quad G_m^+ = e^{-(t_0+\tau)s}.$$

To design the controller is only used the invertible part of the model is the static gain, K. Thus, $G_m^-(s)$ is written as follows:

$$G_m^-(s) = \frac{X^-(s)}{U(s)} = K. \tag{9}$$

Equation (9) is rewritten in the form:

$$X^-(t) = Ku(t) \tag{10}$$

To determine the equivalent part $u_{eq}(t)$ of the SMC, the following sliding surface is proposed:

$$\sigma(t) = \lambda \int e(t)dt + \int e^-(t)dt + C \tag{11}$$

where $e(t) = Ref - X(t)$ and $e^-(t) = Ref - X^-(t)$.

Ref is the reference or set point of the system, $e^-(t)$ is the error between the reference, Ref and the model output without deadtime, $X^-(t)$. $X(t)$ is the controlled variable and $e(t)$ is the error between the reference, Ref and the output of the process.

The role of the arbitrary constant C is to express the classes of all primitive functions different from the given original.

According to the Filippov construction procedure of the dynamic equivalent [23], the sliding condition is given by:

$$\dot{\sigma}(t) = 0 \tag{12}$$

Therefore, the derivative of (11) is:

$$\dot{\sigma}(t) = \lambda e(t) + e^-(t) = 0. \tag{13}$$

Thus substituting (10) into (13), the equivalent part of the controller is given by:

$$u_{eq} = \frac{\lambda e(t) + Ref}{K}. \tag{14}$$

The discontinuous or switching part of the controller $u_d(t)$ is defined as:

$$u_d(t) = K_D sgn(\sigma(t)), \tag{15}$$

where the complete control law is given by:

$$u_{SMC}(t) = \frac{\lambda e(t) + Ref}{K} + K_D sgn(\sigma(t)). \tag{16}$$

3.1   Stability Analysis

To verify the stability of the sliding surface, the Lyapunov stability approach is used. The system is stable if the projection of the trajectories of the system on the sliding surface is stable [10,16].

Theorem 1. *If there exists a candidate function* $V = \frac{\sigma^2(t)}{2}$ *such that* $V > 0$ *and* $\dot{V} < \sigma(t)\dot{\sigma}(t) < 0$, *for all* $t \in \mathbb{R}^+$ $(\dot{V} = \frac{dV}{dt})$ *Then the system is stable.*

Taking into account $e^-(t) = Ref - X^-(t)$ the derivative of the sling surface (11) is written as:

$$\dot{\sigma}(t) = \lambda e(t) + Ref - X^-(t) \tag{17}$$

By substituting (10) into (16), the following results are obtained:

$$X^-(t) = \lambda e(t) + Ref + KK_D sgn(\sigma(t)) \tag{18}$$

Thus, (18) is replaced by (17):

$$\dot{\sigma}(t) = -KK_D sgn(\sigma(t)) \tag{19}$$

Therefore, the reachability condition is given by the following:

$$\dot{\sigma}(t)\sigma(t) = -KK_D sgn(\sigma(t))\sigma(t) = -K_D K |\sigma(t)| \tag{20}$$

Hence, in order to ensure stability, $K_D > 0$ and $K > 0$ must be guaranteed.

Finally, to reduce the chattering effect, the sigmoid smoothing function (6) is added to the control law (16), results:

$$u_{SMC}(t) = \frac{\lambda e(t) + Ref}{K} + K_D \frac{\sigma(t)}{|\sigma(t)| + \delta}. \tag{21}$$

3.2   Parameter Tuning Based on PSO

In this work, the Particle Swarm Optimization (PSO) method is used to adjust the parameters of the proposed controller (λ, K_D, δ). The PSO algorithm is used for robust global optimization and is successfully used to adjust parameters in chemical processes [20]. It is used to optimize the performance of the parameters.

The method of computer search and optimization known as PSO was developed by Eberhart and Kennedy in 1995. Biological systems inspired it. It is based on the social behaviors of groups of animals, including flocks of birds and schools of fish. The PSO allows us to optimize a problem from several simple entities (particles) placed in the search space of some problem. Each evaluates its objective function (F_{obj}) in the current position. Each particle determines its movement throughout the search space according to mathematical rules that take into account the position and velocity of the particles. This movement of each particle is influenced by the history of its current position, as well as the best global positions with some random perturbations as the particles traverse the search space [12, 26].

This article proposes an objective minimization function consisting of an error-dependent component (ISE) and a control action-dependent component (TVu). This function allows one to determine a compromise between error and control action.

$$F_{obj} = \sum_{m=1}^{t_f} \left(e_f(k)\right)^2 + \sum_{m=1}^{t_f} |u_{m+1} - u_m| \tag{22}$$

and subject to $U_m \leq |U_{max}|$. where $m \in \mathbb{N}$, t_f is the total experiment time.

The parameters of the proposed controller were tuned with Matlab 2019b software on an Intel (R), Core (TM) $i7 - 7500U$@2.9 GHz PC on Windows 10. For the configuration of the PSO algorithm, an initial population of 50 particles and a maximum of 70 iterations were considered.

4   Simulation Results

In this section, the performance of the proposed controller is analyzed. This has been tested on a non-linear chemical process with long-time delay. The tests are carried out with a reference change, and external disturbances. In order to design the controller, the approximate FOPDT model is obtained by identifying the reaction curve method [22]. Finally, the proposed control parameters were determined using Particle Swarm Optimization (PSO) as presented in Sect. 3.2.

4.1   Performance Indicators

The performance of the controller is measured using the following indices.

Integral of the Squared Error (ISE): It is calculated by integrating the square of the error of the system over a fixed period. A small value of this quantity indicates small variations in the process response and the setpoint;

$$ISE = \int_0^t e(t)^2 dt \tag{23}$$

Integral of the Total Variation of the Control Signal (TVu): Measures the effort of the control signal. A low value of this amount implies that the control signal varies smoothly, extending the life of the final control element. It can be determined as follows:

$$TVu = \sum_{k=1}^t |u_{k+1} - u_k| \tag{24}$$

Maximum Overshoot (Mp%): It is the maximum response value minus the steady-state value divided by the steady-state value and the result multiplied by 100.

Settling Time (ts): This is the time required for the response to reach a steady state and remain within the specified tolerance bands around the final value.

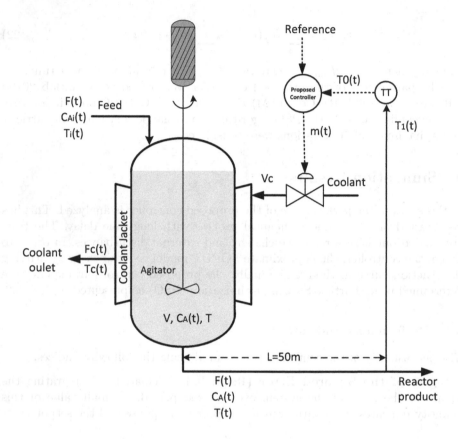

Fig. 2. Chemical Reactor with long-time delay Process.

4.2   Chemical Reactor with Long-Time Delay

Figure 2 illustrate a continuous stirred tank reactor (CSTR) process taken from [9]. The CSTR process presents a long-time delay due to the location of the temperature transmitter. An exothermic reaction $A \rightarrow B$ is carried out in the CSTR process. To reduce the heat of the reaction, an inlet coolant is used, which circulates through the reactor jacket. The control objective is to maintain or change the outlet temperature $T(t)$ within a range by manipulating the position of the valve $m(t)$ by varying the flow $Fc(t)$. Furthermore, the feed temperature $Ti(t)$, the feed flow $F(t)$, and the reactant concentration $C_A(t)$ are considered constant. For practical purposes, the temperature control is calibrated in the range of 80 to 100 °C and the valve is operated between the values of 0 to 1. The Table shows the operating values of the CSTR processes. The mathematical model and all parameters of the CSTR process are presented in [9] (Table 1).

Table 1. Operation parameters

Variable	Value	Variable	Value
C_A	1.113 kgmol/m^3	C_{Ai}	2.88 kgmol/m^3
T	88 °C	$F(t)$	0.45 m^3/min
T_i	66 °C	m	0.254 fraction CO
T_{ci}	27 °C	T_C	50.5 °C
Set point	88 °C	TO	0.4

For the purpose of designing the proposed controller, the dynamics of the CSTR processes is approximated to a FOPDT model using the reaction curve method. Thus, a 10% change in the process input ($m(t)$) is made. In (25) the FOPDT model is presented.

$$G(s) = \frac{1.67}{12.56\,s + 1} e^{-1.67\,s} \tag{25}$$

where the controllability ratio is given by $\left(\frac{t_0}{\tau} \approx 3.4\right)$. Thus, this is considered a process with a long-time delay, which makes it difficult to control.

The proposed control is compared against a PI controller with Smith Predictor (SP) which was tuned using the method proposed by Kaya [11] and against the IM-SMC controller proposed by Camacho et al. [4].

4.3   Reference Change Test

For this test, the process starts with the operating conditions, two reference changes are made. The first a change from 88 to 90 °C at 400 [min] and the second a change from 90 to 94 °C at 1000 [min]. The tuned parameters of the three controllers are presented in Table 2.

Table 2. Controller design parameters for CSTR

Parameter	Proposed	IM-SMC	Kaya - SP
K_p	–	–	1.1976
T_i	–	–	8.9618
λ	0.001	0.0013	–
K_D	0.0419	0.1888	–
δ	0.8894	0.6800	–

The temperature response is shown in Fig. 3, where it can be seen that the controllers take the process towards the reference, however the proposed controller presents the lowest overshoot. In addition the IM-SMC controller presents the highest ts.

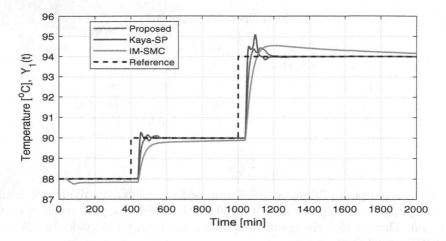

Fig. 3. Response of the temperature reactor $T(t)$ for reference test.

The control actions are shown in Fig. 4, where you can see that the Kaya-SP controller presents peaks in the control actions when the reference changes begin. The control actions are within the operating range.

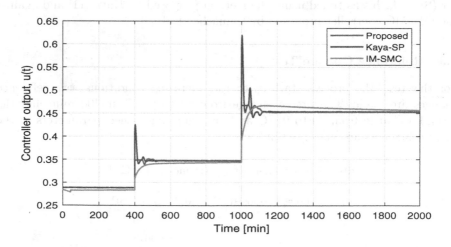

Fig. 4. Controller output $m(t)$ for reference test on CSTR.

In order to have an idea of the global performance of the controllers, a normalized radial chart is presented in Fig. 5, where the indices IAE, ISE, TVu, ts, $Mp\%$ are used. It can be noted that the proposed controller and the Kaya-SP controller present similar performance in terms of IAE, ISE, and ts, however, the proposed controller presents better performance in terms of TVu and $Mp\%$. On the other hand, the IM-SMC controller presents a poor performance in terms of IAE, ISE, and ts.

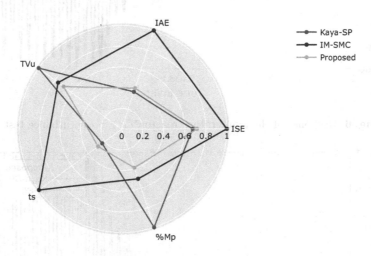

Fig. 5. Normalized radar chart for reference test on CSTR.

4.4 External Disturbance Test

In this test, from the operating conditions, two disturbances in the feed flow $F(t)$ are made. First, it decreases from 0.4 [m^3/min] at 250 [min] and in the second it decreases from 0.4 to 0.35 [m^3/min] at 1000 [min]. For this test, the proposed controller parameters were tuned using PSO, and the following parameters $\lambda = 0.0010$, $K_D = 0.0829$ and $\delta = 0.0010$ were obtained.

The temperature response for the external disturbance test is shown in Fig. 6. Where, it can be noted that only the proposed controller is capable of maintaining the temperature at the operating condition (88 °C) when both external disturbances occur. The Kaya-SP controller is not capable of rejecting the second external disturbance, and the IM-SMC controller has a very high ts.

Figure 7 shows the control actions for the external disturbance test, where it can be seen that for the proposed controller and the IM-SMC controller, their actions are within the operating range. However, for the Kaya-SP controller, its control action after the second external disturbance oscillates within the maximum permitted operating limits.

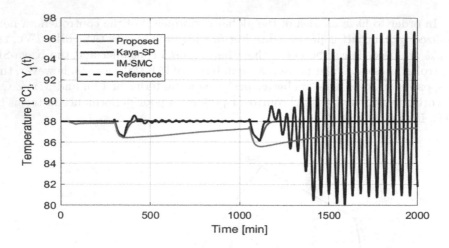

Fig. 6. Response of the temperature reactor $T(t)$ for disturbance test.

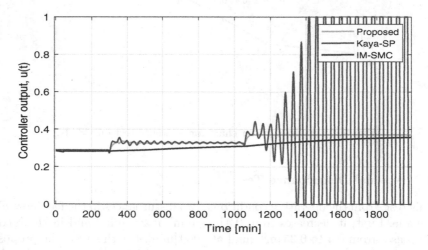

Fig. 7. Controller output $m(t)$ for disturbance test on CSTR.

A normalized radial chart is presented in Fig. 8, which allows us to have a global idea of the performance of the controllers. In this way, it can be noted that the proposed controller and the IM-SMC controller present similar performances in terms of the TVu and ISE indices. On the other hand, the Kaya-SP controller presents poor performance in terms of the IAE, TVu and ISE indices. Finally, as can be seen, the proposed controller presents the best overall performance compared to the other controllers.

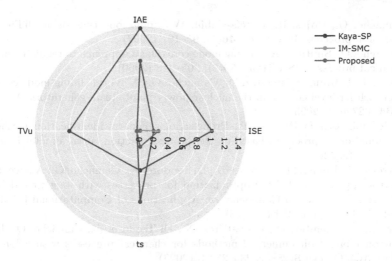

Fig. 8. Normalized radar chart for disturbance test on CSTR.

5 Conclusion

A novel approach was shown. The approach mixed the Smith predictor scheme, the sliding mode control methodology, and the PSO optimization algorithm. The proposed approach considered including the Taylor series approximation to put together a dead-time term as the non-invertible part and a gain for the invertible part; in this way, the controller synthesis is too easy. The PSO algorithm has been used here to fine-tune parameters for the proposed approach, dramatically increasing the efficiency of designing robust controllers for nonlinear systems.

Future work in this research includes finding the appropriate tuning equations for the proposed controller, validating the proposed control strategy in real time, using hardware-in-the-loop (HIL) simulations, and implementing the proposed algorithm in an embedded system.

Acknowledgment. Marco Herrera thanks the Advanced Control Systems Research Group at USFQ for his research internship.

References

1. Åström, K.J., Hägglund, T., Astrom, K.J.: Advanced PID Control, vol. 461. ISA-The Instrumentation, Systems, and Automation Society Research Triangle Park (2006)
2. Báez, E., Bravo, Y., Leica, P., Chávez, D., Camacho, O.: Dynamical sliding mode control for nonlinear systems with variable delay. In: 2017 IEEE 3rd Colombian Conference on Automatic Control (CCAC), pp. 1–6. IEEE (2017)
3. Camacho, O., Leiva, H.: Impulsive semilinear heat equation with delay in control and in state. Asian J. Control **22**(3), 1075–1089 (2020)

4. Camacho, O., Rojas, R., García-Gabín, W.: Some long time delay sliding mode control approaches. ISA Trans. **46**(1), 95–101 (2007)
5. Camacho, O., Smith, C.A.: Sliding mode control: an approach to regulate nonlinear chemical processes. ISA Trans. **39**(2), 205–218 (2000)
6. Espín, J., Estrada, S., Benítez, D., Camacho, O.: A hybrid sliding mode controller approach for level control in the nuclear power plant steam generators. Alex. Eng. J. **64**, 627–644 (2023)
7. Gude, J.J., García Bringas, P.: Proposal of a general identification method for fractional-order processes based on the process reaction curve. Fractal Fractional **6**(9), 526 (2022)
8. Herrera, M., Benítez, D.S., Pérez, N., Di Teodoro, A., Camacho, O.: A novel hybrid control approach with PSO optimization for processes with long time-delay. In: 2023 IEEE Colombian Conference on Applications of Computational Intelligence (ColCACI), pp. 1–6. IEEE (2023)
9. Herrera, M., Benítez, D., Pérez-Pérez, N., Di Teodoro, A., Camacho, O.: Hybrid controller based on numerical methods for chemical processes with a long time delay. ACS Omega **8**(28), 25236–25253 (2023)
10. Herrera, M., Camacho, O., Leiva, H., Smith, C.: An approach of dynamic sliding mode control for chemical processes. J. Process Control **85**, 112–120 (2020)
11. Kaya, I.: Tuning smith predictors using simple formulas derived from optimal responses. Ind. Eng. Chem. Res. **40**(12), 2654–2659 (2001)
12. Kennedy, J., Eberhart, R.: Particle swarm optimization. In: Proceedings of ICNN 1995-International Conference on Neural Networks, vol. 4, pp. 1942–1948. IEEE (1995)
13. Liptak, B.G., et al.: Instrument Engineers' Handbook: Process Control and Optimization, vol. 2. CRC Press (2018)
14. Mejía, C., Salazar, E., Camacho, O.: A comparative experimental evaluation of various smith predictor approaches for a thermal process with large dead time. Alex. Eng. J. **61**(12), 9377–9394 (2022)
15. Ming-Xia, C., Jin-di, Z., Hong, Z.: Research on control algorithms of systems with long time delay. In: Proceedings of the International Symposium on Big Data and Artificial Intelligence, pp. 151–156 (2018)
16. Morales, L., Aguilar, J., Camacho, O., Rosales, A.: An intelligent sliding mode controller based on LAMDA for a class of SISO uncertain systems. Inf. Sci. **567**, 75–99 (2021)
17. Normey-Rico, J.E., Camacho, E.F.: Control of Dead-Time Processes. Advanced Textbooks in Control and Signal Processing, Springer, Heidelberg (2007). https://doi.org/10.1007/978-1-84628-829-6
18. Normey-Rico, J.E., Santos, T.L.M., Flesch, R.C.C., Torrico, B.C.: Control of dead-time process: from the smith predictor to general MIMO dead-time compensators. Front. Control Eng. 19 (2022)
19. O'Dwyer, A.: Handbook of PI and PID Controller Tuning Rules, 3rd edn. Distributed by World Scientific Publishing, Imperial College Press (2009)
20. Revelo, J., Herrera, M., Camacho, O., Alvarez, H.: Nonsquare multivariable chemical processes: a hybrid centralized control proposal. Ind. Eng. Chem. Res. **59**(32), 14410–14422 (2020)
21. Sardella, M.F., Serrano, M.E., Camacho, O., Scaglia, G.J.: Design and application of a linear algebra based controller from a reduced-order model for regulation and tracking of chemical processes under uncertainties. Ind. Eng. Chem. Res. **58**(33), 15222–15231 (2019)

22. Seborg, D.E., Edgar, T.F., Mellichamp, D.A., Doyle, F.J., III.: Process Dynamics and Control. Wiley, Hoboken (2016)
23. Slotine, J.J.E., Li, W., et al.: Applied Nonlinear Control, vol. 199. Prentice Hall, Englewood Cliffs (1991)
24. Tsai, H.H., Fuh, C.C., Ho, J.R., Lin, C.K., Tung, P.C.: Controller design for unstable time-delay systems with unknown transfer functions. Mathematics 10(3), 431 (2022)
25. Utkin, V., Poznyak, A., Orlov, Y.V., Polyakov, A.: Road Map for Sliding Mode Control Design. SM, Springer, Cham (2020). https://doi.org/10.1007/978-3-030-41709-3
26. Yang, X.S.: Nature-Inspired Optimization Algorithms. Academic Press (2020)
27. Zhang, X.M., Han, Q.L.: Time-delay systems and their applications. Int. J. Syst. Sci. 53(12), 2477–2479 (2022)

Feature Selection of the Spectral Signature of Cocoa Bean Based on Visible and Near-Infrared Spectroscopy for Cadmium Estimation

César Cruz, Eduardo Grados, Gerson La Rosa(iD), Juan Valdiviezo(iD), and Juan Soto$^{(\boxtimes)}$ (iD)

Universidad de Piura, Piura, Perú
juan.soto@udep.edu.pe

Abstract. Cocoa production in Peru has grown significantly in the last decade. Peru is the eighth largest producer of cocoa beans and the second largest producer of organic cocoa and fine aroma cocoa in the world. However, the presence of high cadmium content restricts access to international markets. The European Union has established a limit of 0.8 ppm for the presence of heavy metals due to their harmful effects on health. To improve the cocoa bean production process and establish adequate and non-destructive cocoa quality control, this article seeks to estimate the percentage of cadmium in the cocoa bean using its spectral signature. This signature includes the wavelength of 400 nm to 900 nm, covering the visible and near-infrared spectrum. 233 samples from the department of Huánuco, located in the north center of Peru, and the partial least squares (PLS) method with feature selection algorithms is used to identify the wavelengths that most contribute to the estimation of cadmium in cocoa beans. This selection of wavelengths has allowed us to improve the precision of the cadmium estimation, reaching an R^2 of 75.67% and an average error of 0.19 ppm with the test data. The selected wavelengths can be considered to design an automatic system that can be implemented in real conditions.

Keywords: Spectral signature · cadmium in cocoa beans · selecting wavelength

1 Introduction

Peru is the eighth largest producer of cocoa beans and the second largest producer of organic cocoa and fine aroma cocoa in the world. The main importing countries of Peruvian cocoa are Switzerland, France, Holland, Venezuela, United Kingdom, Italy, United States and Germany [1].

In Peru there are more than 100 thousand farmers who cover more than 180 thousand hectares of cocoa fields. The majority of Peru's cocoa production is focused on the center and south of the country where the department of Huánuco is located. The highest levels of cocoa cadmium in Peru come from areas with active mining activity. In [2] defines

© The Author(s), under exclusive license to Springer Nature Switzerland AG 2024
A. D. Orjuela-Cañón et al. (Eds.): ColCACI 2023, CCIS 1865, pp. 72–83, 2024.
https://doi.org/10.1007/978-3-031-48415-5_6

that in terms of the national cocoa production in Peru, 9.2 to 41%might be affected by the Cd restrictions, for the 0.8 and 0.5 ppm thresholds, respectively.

Contamination by cadmium and other heavy metals represents one of the greatest health risks, since high levels of these can cause different diseases. For this reason, the European Union established the maximum permitted concentration of cadmium in cocoa, chocolate and its derivatives at a maximum of 0.8 ppm. This rule came into effect in 2019.

The methods, protocols and equipment used to measure cadmium concentrations in cocoa are indicated in [3]. Atomic absorption spectrometry, atomic absorption spectrometry after microwave digestion and atomic absorption spectrometry after obtaining dry ashes are mainly used in cocoa beans. These methods require adequate sample preparation and selection of appropriate laboratories, implying additional costs for cocoa producers and exporters. Faced with this, there is a need for alternatives to estimate cadmium concentrations in cocoa beans. Maps estimating cadmium concentrations in regions can be found in the literature, however, they are only indicative and there will be uncertainty of the actual cadmium concentrations for specific plots [4]. Alternatives are required that can be implemented in collection areas or production lines to know in real time the presence of cadmium in cocoa beans.

Techniques based on hyperspectral images, which cover the ultraviolet, visible and infrared spectrum, have emerged as reliable methods for rapid, effective, low-cost and non-destructive analysis that allows the characterization and quantification of food quality parameters related to variables that are traditionally analyzed in laboratories, it also allows complex characteristics of fruits to be analyzed such as maturity, defects, mechanical damage and sensory quality, and even the use of these techniques allows us to know the origin of food, identify diseases in agricultural products and determine if there are adulterations. In the purity of food [5–7]. In [8] the optical properties of apple, peach, pear, kiwi, plum, cucumber, zucchini and tomato were determined from the spectrum of the visible region and the near infrared (500–1000 nm) using a diffuse reflectance technique. In [9] the content of sugars, anthocyanins and pH in wine grapes is determined using hyperspectral images in the reflectance mode from 380 to 1028 nm. In [10] use hyperspectral imaging to evaluate the ripening of red-skinned and soft-skinned peaches using a combination of wavelengths close to the chlorophyll absorption peak at 680 nm to discriminate ripening stages. In [11], three optical maturity indices based on chlorophyll content were compared with the objective of non-destructively estimating firmness in "Braeburn" and "Cripps Pink" apples: the absorption coefficient measured at 670 nm, the IAD index and the NDVI index. In [12] a classification of blueberry fruits and leaves was developed based on spectral signatures in the range of 200–2500 nm in the following categories: leaves, ripe fruit, almost ripe fruit, almost young fruit and young fruit.

In cocoa, techniques based on hyperspectral images have been applied to predict the chemical components of cocoa such as the moisture content in the drying process and the fat content of cocoa beans. In [13–18] hyperspectral imaging has been used to relate the cocoa fermentation process, analyzing the chemical changes of the cocoa bean during fermentation such as the PH level, and evaluating different methods including the cut test, colorimetry, hyperspectral indices, fluorescence spectroscopy, NIR spectroscopy, and gas chromatography-mass spectrometry (GC-MS). In [19] a classification system for cocoa beans is established according to their degree of fermentation using the spectral range of 970 to 2500 nm with a spectral sampling of 5.45 nm.

In [20–23], preliminary studies are carried out where the complete spectral signature in the range of 400 nm to 900nm is used to identify prediction models that estimate the level of cadmium in cocoa in Piura, located in the north of Peru, multiple Machine Learning algorithms are identified and compared which have as input the complete hyperspectral signature.

This document presents a strategy that applies Machine Learning algorithms and feature selection algorithms to identify the most important wavelengths of the visible and near-infrared spectrum of cocoa in order to estimate the cadmium content in real time and make the implementation of an automatic online system.

The document is organized as follows: Sect. 2 shows the data used and the management carried out for the estimation of the model, the partial least squares (PLS) model and the calibration procedure used are shown, and the method is also defined. Used for the identification of the most important spectral bands for prediction, in Sect. 3 the results obtained are shown and we compare the generated models, and finally the conclusions of the research are given in Sect. 4.

2   Materials and Methods

2.1   Cocoa Bean Samples Analyzed

233 samples from the department of Huanuco have been analyzed. The analyzed samples have an average of 0.71 ppm cadmium concentration, identifying samples with a maximum value of 2.3 ppm and a minimum value of 0.08 ppm (see Fig. 1). Following European regulations, 34.33% of the study samples exceed 0.8 ppm.

To obtain the hyperspectral image, the Resonon camera model Pika II was used, which has a spectral range of 400 nm to 900 nm and a spectral resolution of 2.1 nm. The hyperspectral cube obtained is made up of 240 planes or images of the same scene, corresponding to their respective wavelengths.

The spectral signature has been obtained from the hyperspectral cube, which represents the reflectance values with respect to the wavelengths. To generate the spectral signature of each sample, a region of interest in the grain known as Region of Interest (ROI) is determined and the average is calculated. Figure 2 shows the spectral signature of the 233 samples analyzed, showing the value of the reflectance achieved in the range of 400 nm to 900 nm, the shaded area shows the reflectance values and the central line the average of the reflectances obtained from the samples.

We will use 75% of the samples to perform the training and calibration of the PLS model and the feature selection algorithms. The remaining 25% of the samples will be used for comparison of the results obtained. The selection is made randomly.

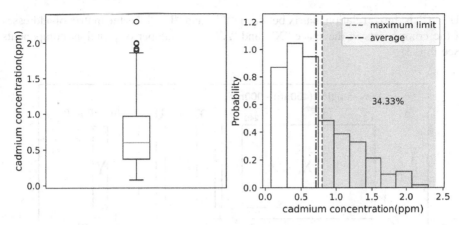

Fig. 1. Cadmium concentration.

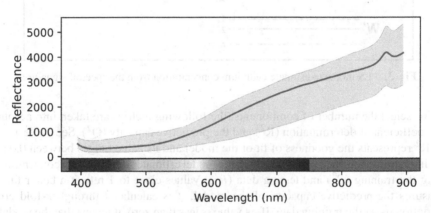

Fig. 2. Spectral signature for cocoa beans.

2.2 Partial Least Squares Regression (PLS)

The PLS model is based on reducing the number of variables in the data set, combining them and generating new latent variables called principal components that will be the directions that maximize the covariance between the reflectance obtained at each of the wavelengths of the spectrum (X) and the cadmium concentration value of the cocoa bean (Y).

$$Y_{train} = X_{train}B_{pls} \tag{1}$$

$$B_{pls} = W\left(P'W\right)^{-1}C' \tag{2}$$

The PLS model is shown in Eq. 1, where the coefficient vector B_{pls} and the number of components will be estimated using the training data. The calculation of B_{pls} is obtained by Eq. 2 using the NIPALS algorithm explained in [24]. "T" and "U" are the score matrix of the space "X" and "Y" respectively, "W" is the correlation matrix between "X" and

"U", "C" is the correlation matrix between "Y"." and "T", "P" is the matrix of addresses of the components in the space "X", and "A" is the number of principal components (See Fig. 3).

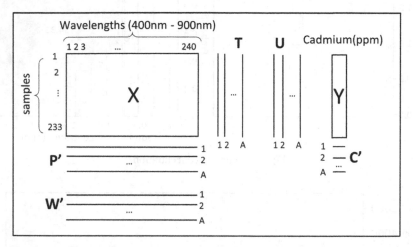

Fig. 3. PLS model to estimate cadmium concentration from the spectral signature.

To select the number of components, the following metrics are taken into account: the coefficient of determination (R^2) and the predictive capacity (Q^2). See Table 1.

R^2 represents the goodness of fit of the model and its value ranges between 0.0 and 1.0, it quantifies the precision with which the model estimates the cadmium concentration using the training (R_t^2) and testing data (R_v^2). Values close to 1 reflect a better fit. Q^2 measures the predictive capacity of the model, it is calculated through k-fold cross validation using the training data. If its value is less than zero, it means that the model is not suitable for prediction; on the contrary, if its value is close to 1 it indicates that the model is relevant for predicting cadmium concentration.

2.3 Variable Importance in Projection (VIP)

The VIP values measure the influence of each independent variable in the PLS model. The VIP values are calculated taking into account the weights of the latent variables of the PLS model, quantifying the contribution to the prediction result. The calculation of the VIP value is explained in detail in [25]. To select the significant variables, those variables whose VIP value is equal to or greater than one will be taken into account, which indicates that they have high relevance for the model.

$$VIP_j = \sqrt{M \, \Sigma_{a=1}^A \left[w_{ak}^2 SCEY(a) \right] / SCEY(total)} \qquad (3)$$

where: "SCEY(a)" is the sum of the squares of the variable "Y" explained by the component "a", "SCEY(total)" is the total sum of the squares of the variable "Y" explained by the model with "A" principal components, "M" is the total wavelengths used in the model, VIP_j is the VIP value for wavelength "j".

2.4 Backward Variable Elimination (BVE)

In the BVE algorithm, the wavelengths are ordered according to the value of B_{pls}. The algorithm starts with the 240 wavelengths and in each iteration the following is done:

- Identify the wavelength that has the smallest absolute value of B_{pls}
- Eliminate the identified wavelength and adjust the PLS model with the rest of the wavelengths (which we will call selected wavelengths).
- Record the value of Q^2 constructed with cross validation (k-fold = 10) and varying the number of principal components.
- Select the PLS model where the number of principal components allows achieving the highest Q^2.
- Update the value B_{pls} obtained from the selected PLS model.
- We store the value of Q^2 and the selected wavelengths in this iteration.

After finishing the algorithm (when only one wavelength remains), the iteration where the highest value of Q^2 was reached is identified.

Table 1. Metrics used in model calibration and evaluation.

Metrics	Variables
$R_t^2 = 1 - \dfrac{\sum_i^n \left(y_i - \hat{y}_{i,t}\right)^2}{\sum_i^n (y_i - \bar{y})^2}$	y_i = cadmium concentration of sample "i" $\bar{y}$ = average value of cadmium concentration $\hat{y}_{i,t}$ = cadmium prediction of sample "i" using a PLS model built with the training data where sample "i" is included
$R_v^2 = 1 - \dfrac{\sum_i^n \left(y_i - \hat{y}_{i,v}\right)^2}{\sum_i^n (y_i - \bar{y})^2}$	$\hat{y}_{i,v}$ = cadmium prediction of sample "i" using a PLS model built with the training data where sample "i" is not included
$Q^2 = 1 - \dfrac{\sum_i^n \left(y_i - \hat{y}_{i,k}\right)^2}{\sum_i^n (y_i - \bar{y})^2}$	$\hat{y}_{i,k}$ = prediction of sample "i" using a PLS model built with a subset of data "k" created by cross-validation where sample "i" is not included

3 Results

3.1 PLS Model Using Full Spectral Signature

Different PLS models of the spectral signature and cadmium concentration of the cocoa bean are carried out, varying the number of principal components: $A = \{1, 2, ..., 50\}$.

For cross validation, the training data is divided into 10 subsets (k = 10), which simulates in each iteration the construction of a PLS model with 90% of the training data and calculating the prediction with 10% of the data. Remaining training sessions.

Figure 4 shows the values of R^2 and Q^2 varying the number of principal components. Based on these results, it has been decided to use 17 principal components, where the value of Q^2 begins to fall noticeably.

The value of R_t^2 y R_v^2 obtained with the spectral signature was 65.71% and 71.26% respectively. In Fig. 5 the results of the predicted cadmium values compared to the actual concentrations using the training data and using the testing data are shown.

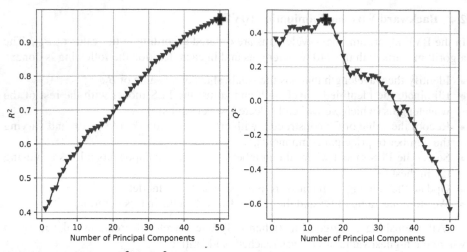

Fig. 4. Values of R^2 and Q^2 of the PLS models built with the full spectral signature.

3.2 PLS Model Using VIP Metric (PLS-VIP)

Table 2 shows the lengths for which VIP values greater than 1 were obtained; these wavelengths have been grouped into 5 zones. A total of 122 wavelengths were obtained.

With the 5 zones obtained, a new PLS model has been built, following the same procedure as in point 3.2. The Q^2 values have been obtained and it was found that the number of components with the best performance is 23 components. The value of R_t^2 and R_v^2 obtained with the PLS-VIP model was 73.87% and 72.39% respectively.

3.3 PLS Model Using BVE Algorithm (PLS-BVE)

Table 3 shows the selected lengths after applying the BVE algorithm. A total of 46 wave-lengths were obtained.

With the selected wavelengths, a new PLS model has been built, following the same procedure as point 3.2. The Q^2 values have been obtained and it was found that the number of components with the best performance is 11 components. The value of R_t^2 and R_v^2 obtained with the PLS-VIP model was 72.31% and 75.67% respectively.

Table 2. Selected wavelengths by PLS-VIP model.

Description	Upper wavelength (nm)	Lower wavelength (nm)
Zone 1	394.35	400.61
Zone 2	534.01	665.71
Zone 3	713.72	790.95
Zone 4	811.83	822.26
Zone 5	874.45	893.23

Table 3. Selected wavelengths by PLS-BVE model.

Selected wavelengths (nm)				
398.52	519.59	805.57	838.97	882.80
402.70	569.69	807.66	841.05	884.89
404.79	651.10	809.74	843.14	886.98
408.96	653.19	816.00	845.23	889.06
411.05	667.80	818.09	847.32	891.15
421.49	669.89	820.18	849.40	893.24
429.84	747.12	822.27	851.49	
438.19	749.21	824.35	853.58	
440.27	751.30	826.44	855.67	
454.88	803.48	836.88	857.75	

3.4 Discussion of Results

Table 4 shows a summary of the metrics of interest on the performance of the models developed in this research.

The best results were achieved with the PLS-VIP model where it was possible to reduce the number of wavelengths by half. However, the PLS-BVE model has also allowed us to obtain acceptable results.

In Fig. 5 is observed that at a graphic level the results are very similar if we compare the real value and the predicted value of the cadmium concentration using the PLS, PLS-VIP and PLS-BVE models.

The bands selected between the PLS-VIP model and the PLS-BVE model do not coincide for the most part (See Fig. 6). This is also observed in the prediction errors graph (See Fig. 7).

Table 4. Summary of results.

Model	Number of selected wavelengths	R_t^2	R_v^2	Minimum deviation (ppm)	Maximum deviation (ppm)	Average deviation (ppm)
PLS	240	65.71%	71.26%	0.008242	0.850392	0.198659
PLS-VIP	122	72.31%	75.67%	0.002075	0.720758	0.190822
PLS-BVE	46	63.35%	72.86%	0.000112	0.890005	0.191385

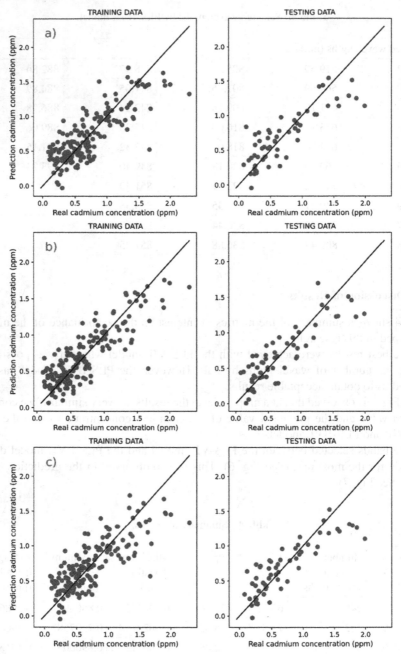

Fig. 5. Predicted value and actual value of cadmium concentration for training and testing data using PLS (a), PLS-VIP(b), PLS-BVE(c).

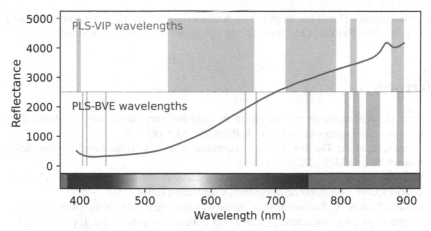

Fig. 6. Wavelengths selected using PLS-VIP (upper bars) and PLS-BVE (lower bars).

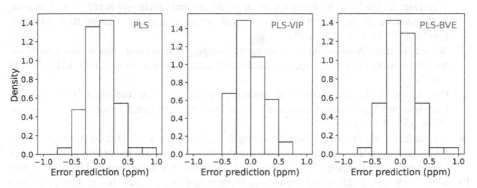

Fig. 7. Model prediction error for testing data.

4 Conclusions

The results of the research show that it is possible to obtain an estimate of the cadmium content in the cocoa bean using hyperspectral vision techniques. A PLS model was found that predicts the cadmium content values with an acceptable error. The inputs to the model is the reflectance obtained from the hyperspectral images.

From the PLS model with the complete signature, taking the reflectance of the 240 wavelengths of the spectral signature, an R^2 value of 71.26% was obtained with the test data, which indicates that the prediction model is good.

Through the variable selection algorithms, it has been possible to improve the PLS model, obtaining a PLS-VIP model that uses 122 wavelengths and reaches an R^2 of 75.67 and a PLS-BVE model that uses 46 wavelengths and reaches a R^2 of 72.86% using test data.

The results achieved with the selected wavelength proposals can be taken as a reference for the design of an automatic system based on hyperspectral vision technique that can be implemented in real conditions.

Acknowledgment. Thank the Biodiversity Institute for the samples of cocoa beans and the Automatic Control Systems Laboratory for the equipment and facilities provided.

References

1. Viera, G.: Aplicación de procesamiento de imágenes para clasificación de granos de cacao según su color interno. Universidad de Piura, Piura (2018)
2. Thomas, E., et al.: The distribution of cadmium in soil and cacao beans in Peru. Sci. Total Environ. **881**, 163372 (2023)
3. Pastás, K., Cacuango, L., Vasco, E.: Selección del laboratorio más idóneo para el análisis de cadmio, Quito: Ministerio de Agricultura y Ganadería (2021)
4. Vanderschueren, R., Pulleman, M.: Cadmio en cacao: de dónde viene, cómo se regula y por qué preocupa a los productores, Clima-LoCa Políticas en Síntesis, no. 1, p. 10 (2021)
5. Eh Teet, S., Hashim, N.: Recent advances of application of optical imaging techniques for disease detection in fruits and vegetables: a review. Food Control **152**, 109849 (2023)
6. Wen-Hao, S., Da-Wen, S.: Fourier transform infrared and raman and hyperspectral imaging techniques for quality determinations of powdery foods: a review. Compr. Rev. Food Sci. Food Saf. **17**(1), 104–122 (2018)
7. Wieme, J., et al.: Application of hyperspectral imaging systems and artificial intelligence for quality assessment of fruit, vegetables and mushrooms: a review. Biosys. Eng. **222**, 156–176 (2022)
8. Qin, J., Lu, R.: Measurement of the optical properties of fruits and vegetables using spatially resolved hyperspectral diffuse reflectance imaging technique. Postharvest Biol. Technol. **49**(3), 355–365 (2008)
9. Fernandes, A.M., Franco, C., Mendes-Ferreira, A., Mendes-Faia, A., Leal da Costa, P., Melo-Pinto, P.: Brix, pH and anthocyanin content determination in whole Port wine grape berries by hyperspectral imaging and neural networks. Comput. Electron. Agric. **115**, 88–96 (2015)
10. Lleó, L., Roger, J., Herrero-Langreo, A., Diezma-Iglesias, B., Barreiro, P.: Comparison of multispectral indexes extracted from hyperspectral images for the assessment of fruit ripening. J. Food Eng. **104**, 612–620 (2011)
11. Zanella, A., et al.: Correlating optical maturity indices and firmness in stored "Braeburn" and "Cripps Pink" apples. Acta Horticulturae **1012**, 1173–1180 (2013)
12. Yang, C., Won Suk, L., Williamson, J.G.: Classification of blueberry fruit and leaves based on spectral signatures. Biosyst. Eng. **113**, 351–362 (2012)
13. Mundaca Vidarte, G.A.: Análisis de la calidad del grano de cacao mediante imágenes hiperespectrales usando técnicas de visión artificial, Universidad de Piura, Piura (2016)
14. Zulfahrizal, Z., Meilina, H., Munawar, A.A.: The use of spectral imaging technology combined with support vector machine as a fast and novel tools for fat content determination of intact cocoa beans. In: International Conference on Technology, Innovation, and Society (ICTIS) (2016)
15. Caporaso, N., Whitworth, M.B., Fisk, I.D.: Total lipid prediction in single intact cocoa beans by hyperspectral chemical imaging. Food Chem. **344**, 128663 (2021)
16. Aculey, P.C., et al.: Ghanaian cocoa bean fermentation characterized by spectroscopic and chromatographic methods and chemometrics. J. Food Sci. **75**(6), S300–S307 (2010)
17. Soto, J., Paiva, E., Ipanaqué, W., Reyes, J., Espinoza, D., Mendoza, D.: Cocoa bean quality assessment by using hyperspectral index for determining the state of fermentation with a non-destructive analysis. In: 2017 CHILEAN Conference on Electrical, Electronics Engineering, Information and Communication Technologies, Pucon, Chile (2017)

18. Ruiz Reyes, J.M.: Estudio de la visión hiperespectral en el proceso de fermentación del cacao, Universidad de Piura, Piura (2016)
19. Liu, N., Gonzalez, J.M., Ottestad, S., Hernandez, J.: Application of hyperspectral imaging for cocoa bean grading with machine learning approaches. In: Proceedings of SPIE - The International Society for Optical Engineering, Birmingham (2023)
20. Peltroche Saavedra, G.: Diseño e implementación de algoritmos inteligentes basados en aprendizaje de máquina para la detección de cadmio en granos de cacao mediante imágenes hiperespectrales, Universidad de Piura, Piura (2021)
21. Neyra Hau Yon, J.L.: Determinación en tiempo real de presencia de cadmio en cultivo de cacao aplicando Machine Learning, Universidad de Piura, Piura (2021)
22. Checa Roman, K.V.: Determinación del contenido de cadmio en granos de cacao mediante la aplicación de redes neuronales e imágenes hiperespectrales, Universidad de Piura, Piura (2022)
23. Checa, K., Gamarra, M., Soto, J., Ipanaqué, W., La Rosa, G.: Preliminary study of the relation between the content of cadmium and the hyperspectral signature of organic cocoa beans. In: 2019 IEEE CHILEAN Conference on Electrical, Electronics Engineering, Information and Communication Technologies (CHILECON), Valparaiso, Chile (2019)
24. Höskuldsson, A.: PLS regression methods. J. Chemom. **2**(3), 211–228 (1988)
25. Mehmood, T., Hovde Liland, K., Snipen, L., Sæbø, S.: A review of variable selection methods in Partial Least Squares Regression. Chemometr. Intell. Laboratory Syst. **118**(15), 62–69 (2012)

On the Use of Deep Learning Models for Automatic Animal Classification of Native Species in the Amazon

María-José Zurita[1], Daniel Riofrío[1], Noel Pérez-Pérez[1], David Romo[2], Diego S. Benítez[1]([✉]), Ricardo Flores Moyano[1], Felipe Grijalva[1], and Maria Baldeon-Calisto[1,3]

[1] Colegio de Ciencias e Ingenierías "El Politécnico", Universidad San Francisco de Quito USFQ, Quito 170157, Ecuador
mzuritam@alumni.usfq.edu.ec,
{driofrioa,nperez,dbenitez,rflores,fgrijalva,mbaldeonc}@usfq.edu.ec
[2] Colegio de Ciencias Biológicas y Ambientales "COCIBA",
Universidad San Francisco de Quito USFQ, Quito 170157, Ecuador
dromo@usfq.edu.ec
[3] Ingeniería Industrial, CATENA-USFQ, Universidad San Francisco de Quito USFQ,
Quito 170157, Ecuador

Abstract. Camera trap image analysis, although critical for habitat and species conservation, is often a manual, time-consuming, and expensive task. Thus, automating this process would allow large-scale research on biodiversity hotspots of large conspicuous mammals and bird species. This paper explores the use of deep learning species-level object detection and classification models for this task, using two state-of-the-art architectures, YOLOv5 and Faster R-CNN, for two species: *white-lipped peccary* and *collared peccary*. The dataset contains 7,733 images obtained after data augmentation from the Tiputini Biodiversity Station. The models were trained in 70% of the dataset, validated in 20%, and tested in 10% of the available data. The Faster R-CNN model achieved an average mAP (Mean Average Precision) of 0.26 at a 0.5 Intersection Over Union (IoU) threshold and 0.114 at a 0.5 to 0.95 IoU threshold, which is comparable with the original results of Faster R-CNN on the MS COCO dataset. Whereas, YOLOv5 achieved an average mAP of 0.5525 at a 0.5 IoU threshold, while its average mAP at a 0.5 to 0.95 IoU threshold is 0.37997. Therefore, the YOLOv5 model was shown to be more robust, having lower losses and a higher overall mAP value than Faster-RCNN and YOLOv5 trained on the MS COCO dataset. This is one of the first steps towards developing an automated camera trap analysis tool, allowing a large-scale analysis of population and habitat trends to benefit their conservation. The results suggest that hyperparameter fine-tuning would improve our models and allow us to extend this tool to other native species.

Keywords: Conservation · Camera traps · Deep learning · Faster R-CNN · YOLOv5

© The Author(s), under exclusive license to Springer Nature Switzerland AG 2024
A. D. Orjuela-Cañón et al. (Eds.): ColCACI 2023, CCIS 1865, pp. 84–103, 2024.
https://doi.org/10.1007/978-3-031-48415-5_7

1 Introduction

Detailed, up-to-date and accurate information is needed on the location and behavior of animals for their study and conservation. Considering human threats and their impact on natural habitats, it is increasingly important to monitor the behavior and trends of animal species. Camera traps are an efficient method for this task, as they allow permanent sampling in remote areas in a discrete manner and at low cost [25, 31]. Camera traps are also used to locate endangered species, identify important habitats, and monitor areas of interest. However, extracting information from captured images is usually done manually, which is expensive and time-consuming [25]. Many deep learning models have been proposed for computer vision problems. However, its practical use for wildlife monitoring is limited, mainly due to the complexity of this technology and its high computing requirements [10]. Furthermore, applying models trained in a specific region to images collected in a different geographical area has also been a challenge due to background changes and the presence of previously unseen species. Furthermore, about 70% of the images taken do not contain animals due to a high rate of false triggers [3]. Consequently, to analyze these images, it is necessary to manually filter, count, and classify them by an expert who can identify the species present in each image.

In this regard, Yasuní National Park and Reserve (YNP) is one of the largest reserves in the Amazon basin and has been probed to be in a region of the highest diversity per square kilometer [2]. Like many parks, YNP has several types of conflicts. The most critical problem is oil exploitation and development, which has been ongoing since 1989. The last two development sites (oil blocks) are Blocks 31 and 45, known as ITT. Furthermore, YNP is inhabited by the Kichwa and Waorani people. Within the Waorani, a small fraction of their population remains in voluntary isolation. In the buffer zones of the park, which is part of the transition zone of the Biosphere reserve, there are indigenous and colono peoples who often illegally exploit resources such as timber and wild meat [12, 29]. The total area of the park covers more than one million acres. The park has five guard posts and, at most, 35 park rangers who work 21-7 day shifts. The use of camera traps is being used to monitor the effects of people and oil development within the protected area. For national and international conservation efforts, it is essential to determine the impact of different types of human activity on the presence of wildlife, especially large animals that are considered part of the landscape, as well as some considered keystone species, such as jaguars.

During the past decade, the Tiputini Biodiversity Station (TBS), which belongs to Universidad San Francisco de Quito USFQ, a remote research center in the YNP, has conducted a camera trap project that produces over 700,000 photos and videos of approximately 70 wild species, some endangered or rare. However, the information collected has not yet been used. To this day, there are more than 180,000 images in this database that have been manually processed. TBS has also access to a large data base of crude photos collected by a camera trap project conducted by YNP. Labeling such a vast collection is complex, and the task becomes increasingly more challenging as the database grows. Additionally, it is challenging to recognize independent events: whether a series of images corresponds to the same animal or different ones. Additionally, camera traps are now being used in lodges, indigenous communities, oil operations, and the National Park, resulting in a tremendous increase in data. Automating this process would allow for a large-scale analysis that would ease the manual workload of experts [15].

1.1   Related Works

In terms of related work, there have been several attempts to use Convolutional Neural Networks (CNN) models to automate camera trap image analysis, through object detection tasks. For example, RetinaNet and Region-based CNN (Faster R-CNN) were used in a Unmanned Aircraft Systems (UAS) dataset containing 23,748 images from wild animals on the eastern Tibetan Plateau. The experiments demonstrated that Faster R-CNN had a better performance than RetinaNet [21]. Faster R-CNN has also been used to detect and classify individual pigs, to know how much food they would consume daily, and to optimize their breeding process [35]. There was also a comparison between three CNN-based state-of-the-art object detection CNN-based algorithms: YOLO v3 (You Only Look Once), SSD (Single Shot Detector), and Faster R-CNN using the open source Microsoft COCO dataset [28]. This study concluded that YOLO v3 performs better and is better suited for real-time video analysis. At the same time, Faster R-CNN works well with a small dataset that does not require speed in its analysis, and SSD has a good balance between speed and accuracy. Furthermore, YOLO v5 was implemented for species-level object detection and classification in a temperate Polish forest, Bialowieza. This was the first time this architecture was used for automated mammal recognition using camera trap images. It achieved an average accuracy of 85% F1-score for identifying the 12 most common mammal species in the forest, with a total of 2,659 images with animals [10]. YOLO v5 was also used to identify individual feral cats on an unbalanced small dataset [36]. In another study [26], YOLO v5 was applied for automated pest detection in protected forests, with a precision rate of 97%, as a lightweight and efficient method for devices with IoT.

The usability of a pre-trained Faster R-CNN + InceptionResNet v2 model has also been tested. It was applied to ten different species of wild mammal in color and black & white images, with an accuracy rate of 93% in classification. The authors concluded that this rate could improve with specific training in European mammal species [8]. Another study detected and classified more than 15 animal species using the R-CNN architecture with two different backbones for training: ResNet-101 and InceptionResNet v2; and Inception v3 for classification. Faster R-CNN and YOLO were compared in terms of their performance for the identification, quantification, and localization of desert bighorn sheep, using camera trap images [32]. Finally, Cheema and Anand [9] successfully used Faster R-CNN as an object detection framework to detect animal individuals of patterned species, such as zebras, tigers, and jaguars.

Despite these developments, the current state-of-the-art of these technologies spans wildlife species from North America, Asia, and Africa. Therefore, there is an excellent opportunity to expand this knowledge to the unique species found in the Ecuadorian Amazon rainforest. For example, a threatened species of interest in the YNP is the jaguar (*Panthera onca*), the largest predator in Central and South America. Although these species are difficult to study, as they have low densities, large home ranges, elusive nature, and frequent nocturnal behavior, camera traps have proven to be an efficient technique for this task [4]. Camera traps can be used to examine the availability of jaguar prey, which allows a better understanding of jaguar foraging strategies [34], evaluating the occurrence of prey, and estimating the density of the jaguar. These types of analysis are essential for the conservation and management of these wildlife populations, so significant effort should be invested in managing jaguar prey as a jaguar conservation effort. White-lipped peccary (*Tayassu pecari*) and collared peccary (*Dicotyles tajacu*) are primary targets for hunters and one of the main prey for jaguars. White-lipped peccary populations are threatened by hunting and habitat loss [12].

Therefore, in this paper, we propose a species-level object detection and classification model based on deep learning for white-lipped and collared peccary species. Trained specifically with camera trap images taken from 2004 to 2011 in the Tiputini Biodiversity Station, we benchmark the Faster R-CNN and YOLOv5 models. This work serves as a proof of concept towards the development of an automated native animal classification tool in the near future which may allow for a large-scale analysis of population and habitat trends to benefit their conservation.

2   Materials and Methods

2.1   Database

A dataset comprising of 3,233 images from the TBS camera trap database, containing white-lipped or collared peccaries, was initially selected for preprocessing. Instances from the classes "Taypec" (white-lipped peccary) and "Taytaj" (collared peccary) were labeled by experts in all images. The labels closely enclosed each object, leaving no space between the objects and their bounding boxes. On average, there are 2.3 annotations per image across the two classes. Images of different times of day, weather, lighting, and angles were also used to ensure diversity. Background images have also been added to the dataset to reduce false positives. To prevent bleed from the train/test, duplicate images were automatically removed using the Roboflow AI tool. The average image size was 2.27 megapixels and the median image ratio was 1840 × 1232.

Data Processing. Roboflow's preprocessing tools were used to apply image transformations to all images. Auto-orient was used to strip the images of their exchangeable image file format (EXIF) data and standardize pixel ordering. The images were also resized to a dimension of 416 × 416. Data augmentation techniques were also used to allow the model to generalize better through multiple variations of each source image. Several bounding-box level augmentation techniques have also been used, which alter the content within each bounding box. Horizontal and vertical flips (at image and bounding-box levels) were added to help the model be less sensitive to subject orientation. The random crop technique (at the image and bounding-box levels) was also used to create a random subset of each image. The latter increases resilience to subject translations and camera position, which helps the model to recognize animals that may not always be entirely in a frame or at the same distance from the camera, which is the case for most of the camera trap images from the database. Grayscale augmentation was also used because it increases training variance but does not eliminate color information during testing. Hue augmentation is used as well, as it randomly alters the color channels of each image, helping the model to consider alternative color schemes for objects and scenes, which is helpful in this dataset as the animals and their backgrounds change colors depending on the time of the day, weather conditions, and camera's hardware state. Due to this, saturation augmentation was also applied by adjusting how vibrant each image is. Brightness and exposure augmentation techniques (at image and bounding-box levels) were used to help the model be more resilient to changes in lighting and camera settings. Several camera trap images show a change in camera focus, so a random Gaussian blur augmentation was also used (at image and bounding-box levels) to help the model be more resilient to camera focus. There were many images that contained animals located behind other objects, so cutout augmentation helps the model detect these better by adding randomly generated black boxes on top of the images and, by doing so, encouraging the model to learn more distinguishing features about each class of object. Additionally, bounding-level noise variations help the model to be more resilient to camera artifacts. The final dataset contained 7,733 images after applying the processing steps described above.

2.2   Deep Learning Models

For this proof-of-concept, we decided to use and compare two state-of-the-art deep learning-based detectors. One model is a region proposal-based detector, Faster R-CNN, and the other a regression/classification-based detector, You Only Look Once (YOLO) v5 [1].

The architecture of the R-CNN model has three components: region proposal, deep CNN-based feature extraction, and classification/localization [1]. R-CNN extracts region proposals using selective search, resizing all the extracted crops and passing them through a network, which assigns a category. A selective search algorithm groups regions together based on pixel intensities. Regions with a minimum of 0.5 Intersection Over Union (IoU) (the intersection area between predicted and ground truth boxes, divided by their union area) [22] are labeled, and those with an IoU less than 0.3 are considered background. In the bounding-box regression, CNN predicts the parameters of the bounding box (position and size) [13]. Faster R-CNN uses the same process as an R-CNN, but uses another CNN, the Region Proposal Network (RPN), to generate region proposals, and the Fast R-CNN as a detector network, consisting of a CNN backbone, a Region of Interest (ROI) pooling layer, fully connected layers, and two sibling branches of classification and bounding box regression. The Fast R-CNN framework generally consists of a pretrained CNN and an ROI pooling layer. On the other hand, Faster R-CNN shares the full convolutional features of the image with RPN, which can predict object bounding box and class confidence scores simultaneously [1]. The Faster R-CNN network architecture is shown in Fig. 1 [11].

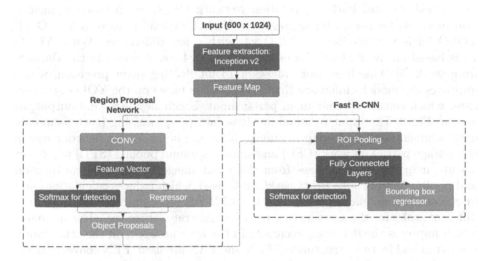

Fig. 1. Faster R-CNN's architecture

While the region-proposal-based framework includes various correlated phases and, therefore, time spent handling different components, regression/classification object detectors reduce this time, as they are a single-stage framework based on class probabilities, mapping directly from image pixel to coordinates and global regression/classification. YOLO predicts the bounding box that uses the top-most feature map and evaluates class probabilities directly. The idea behind this algorithm is to divide the image into $S \times S$ grid cells, and each cell is responsible for predicting the center of the object in the grid cell [1]. Each cell will predict the bounding boxes B and a confidence score for each, which is the Intersection Over Union (IOU).

The idea behind this algorithm is to divide the image into $S \times S$ grid cells, and each cell is responsible for predicting the center of the object in the grid cell [1]. Each cell will predict B bounding boxes and a confidence score for each, which is the IOU.

$$IoU = \frac{\text{Area of Overlap}}{\text{Area of Union}} \qquad (1)$$

Although a cell can predict various bounding boxes and confidence scores, it can only predict one class. Each prediction will have a shape $C + B * 5$, where C is the number of classes, B is the number of predicted bounding boxes, and 5 is the number of elements in each box (x, y, width, height, and confidence). Therefore, for the $S \times S$ matrix, the shape will be $S \times S \times (C + B * 5)$ [23]. YOLO's architecture has three components, head, neck, and backbone, which work to first extract the image visual features and then classify and limit them. The backbone includes convolutional layers that detect key features in an image and process them. The neck uses the features of the convolution layers with fully connected layers, in order to predict bounding-box probabilities and coordinates. The head is the final output layer [23]. YOLO v4 uses CSPDarknet53 as a backbone and Path Aggregation Network (PAN, an instance segmentation method) for parameter aggregation. YOLO v3 head is used in YOLO v4. YOLO v5 is very similar to YOLO v4, with a few differences. While YOLO v5 is based on the PyTorch framework, YOLO v4 was released in the Darknet framework [5]. This framework allows a 16 bit floating point precision, which improves the model's inference time. YOLO v5 is based on the YOLO architecture, which consists of four main parts: input, backbone, neck, and output, as seen in Fig. 2. The input terminal involves data pre-processing (such as mosaic data augmentation and adaptive image filling). The backbone network uses a cross-stage partial network (CSP) and spatial pyramid pooling (SPP) to extract feature maps of different sizes from the input image by multiple convolution and pooling. Specifically, BottleneckCSP is used, which reduces calculations and increases the inference speed, while the SPP structure extracts features from different scales for the same feature map and generates three-scale feature maps, which improves the detection accuracy. In the neck network, on the other hand, the pyramidal feature structures of FPN and PAN are used. FPN conveys strong semantic features from the top feature maps into the lower feature maps, while PAN conveys strong localization features from the lower feature maps to the higher feature maps. The head is the final detection step, used to predict targets of different sizes on feature maps [18]. The structure of the YOLO v5 network is shown in Fig. 2. Glenn et al. in [17] list several recommendations for best performance for YOLOv5, i.e., at least 1,500 images per class and at least 10,000 labeled objects per class.

2.3 Experimental Setup

Training and Test Sets. Roboflow AI tool was used to divide the data set into a training set (6,800 images), a test set (340 images), and a validation set (643 images). As a general rule of thumb, 70% of the dataset was allocated to the training set, 20% to the validation set, and 10% to the test set. This allocation ratio allows the training set to have enough data to learn, the validation set to tune the model, and the testing set to inform its final accuracy.

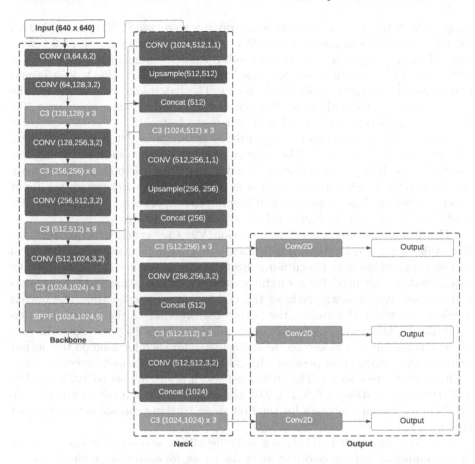

Fig. 2. YOLO v5's architecture

Models Configuration. A model for each object detector type was trained and evaluated for the two classes: Taypec and Taytaj. The models are then compared in terms of performance to set a baseline for hyperparameter optimization in the future. The Faster R-CNN input was resized to a minimum of 600 px for its shortest side and a maximum of 1024 for its larger side, keeping the aspect ratio. The feature extractor used was Inception V2, using batch normalization and producing an output stride of 16px. A bilinear interpolation-based technique was used for the region-of-interest (ROI) pooling layer. For Max-Pooling, the kernel size was set to 2. The batch size used was one and the learning rate was constantly 0.0002. A stochastic gradient descent (SGD) momentum optimizer was used as well. A ReLU activation function was used for the hidden layers, and a Softmax function was used for the output layer. The gradient clipping method used the norm as a clipping threshold. Random horizontal flip was used in addition to those data augmentation techniques described in the preprocessing stage. The total steps for the model's training were 180K. For the YOLO v5 model, an initial learning rate of 0.001 and a final learning rate of 0.1 were applied for training using a once-cycle policy. An SGD momentum optimizer was also used. The activation functions used were Leaky ReLU (for hidden layers) and Sigmoid (for the final detection layer). The batch size was the largest the hardware allows, 64, and the total epochs for the training stage were set to 300. YOLOv5s architecture was used, with 191 layers in total.

The hyperparameters for the Faster RCNN model were determined as follows. The number of classes is 2. The number of steps for the model's training was 180K. For the RPN, the input image is fed to the CNN backbone. The minimum size constraint is 600 and the maximum is 1024. The model resizes the input image satisfying these constrains and keeping the aspect ratio (the shortest side is at least 600px and the longer side does not exceed 1024px).

The feature extractor used is Inception V2, the second generation of the Inception CNN architecture, which uses batch normalization, removing local response normalization. Batch normalization allows the use of higher learning rates, reduces the need for a careful parameter initialization, acts as regularization and, consequently, reduces the training steps needed [16]. The output stride is 16, which determines the output feature size, as the consecutive pixels in the backbone output feature correspond to points 16 pixels apart in each input image. Anchors placed on the input image for each location on the output feature map, which show possible objects in different sizes and aspect ratios at each location, are also set. The anchor generator scales are set to [0.25, 0.5, 1.0, 2.0], the aspect ratios to [0.5, 1.0, 2.0], the height stride and the width stride to 16. The number of proposals for the first stage (region proposal network) is set to 300.

Non Max Supression (NMS), a greedy algorithm, is used to reduce the number of bounding boxes each object gives rise to, as, for each class, it checks for the IoU value between all the bounding boxes and, according to an IoU threshold, determines which refer to the same object and discards the lowest confidence score box. NMS loops over all classes [6]. The score threshold for NMS for the RPN is set to 0, as recommended for Faster R-CNN. Boxes with a score lower than this number are suppressed. The IoU threshold for NMS on the boxes predicted by the RPN is set to 0.7.

For the Region of Interest (ROI) pooling layer, ROI Align, a bilinear interpolation-based technique, is used to crop a patch from a feature map based on a region proposal and resize it to extract a small feature map from each ROI [7] [14]. The initial output size of the bilinear interpolation is set to 14. Additionally, the kernel size (pool size or filter size) of the max pooling operation on the cropped feature map during ROI pooling is set to 2 with a stride of 2.

For the Second Stage Box Predictor, the dropout regularization technique is not used, so the value for the dropout keep probability in a hidden layer is set to 1.0. This means that the probability that the output of the layer is retained is 1.0. For training, the batch size is set to 1, since in one step the model feeds one image at a time. The momentum optimizer is used with a value of 0.9.

This is a method that helps to accelerate the SGD in the relevant direction and dampens the oscillations. It increases for dimensions whose gradients point in the same directions and reduces updates for dimensions whose gradients change directions, gaining faster convergence and reduced oscillation [24]. The learning rate was set to 0.0002, which is the original training learning rate of the faster RCNN Inception V2 model.

Gradient clipping was also used to clip the error derivative to a threshold during backward propagation, to avoid exploding gradients. The gradient clipping-by-norm variable is used, with a threshold of 10.0. This means that the gradients are clipped multiplying the unit vector of the gradients with the threshold. Random horizontal flip, which randomly flips each image with a chance of 50%, was also used as a data augmentation option.

On the other hand, for the YOLO v5 model, the initial learning rate is 0.01 and the final learning rate is 0.1. The once cycle policy for learning rates is applied. This means that in one learning cycle (epoch) there are 2 steps of equal length where in the first step one goes from a lower learning rate to the higher learning rate and then back to the lower learning rate in Step 2. In the last iterations, the learning rate was set below the lower learning rate value [27]. This helps the speed of training and the discovery of the maximum practical learning rate. The SGD momentum was set to 0.937. The warm-up epochs were set to 3, which are the number of epochs in which the model will adjust to the dataset. The IoU threshold was set to 0.2. Several augmentations are also used in the model, such as image mosaic (1.0 probability), image horizontal flip (0.5), image scale (± 0.5 gain), image translation (0.1), image HSV-Hue augmentation (0.015), image HSV-Saturation augmentation (0.7), and image HSV-Value augmentation (0.4). When the training phase was run, the input image size was set to 640×640, the batch size was set to 64, the number of training epochs was set to 300 and the weights are initialized from pre-trained weights, which is recommended for small to medium size datasets. Then, the accuracy of the Yolo model was validated over the testing set, with a batch size of 64. The model was finally used for detection, using the testing set, setting the confidence threshold to 0.6 and using the augmented inference parameter (test-time augmentations or TTA), in order to improve the accuracy.

Assessment Metrics

The performance of the models was evaluated using the Mean Squared Error (MSE) and Cross-Entropy (CE) loss metrics. For YOLOv5 the loss function is composed of the bounding box regression loss (MSE), the objectness loss or confidence of the object presence (Binary Cross Entropy), and the classification loss (Cross-Entropy). The bounding box regression (localization loss) is the loss due to a box prediction that does not cover an object, objectness is the loss due to a wrong box-object IoU prediction, and classification is the loss due to deviations from predicting "1" for the correct classes and "0" for all the other classes for the object in the box. These loss functions are computed for the training set and the validation set. In this case, the results of the latter will be discussed. The faster RCNN loss function uses the same losses: classification, objectness, and localization. However, it calculates the localization loss for both the RPN and the Box Classifier. The precision (P) and the recall (R) were also calculated, and the mean average precision (mAP) was also used as a metric.

3   Results and Discussion

As seen in Figs. 3 and 4, YOLOv5 shows a better mAP in classes averaged over an IoU threshold of 0.5 to 0.95 than Faster-RCNN, suggesting that it has better accuracy and is a more robust model. The graph shows that at an early stage (less than 20 epochs), the value increases significantly and then fluctuates. At approximately 130 epochs, the fluctuation decreases and an overall convergence effect is observed. On the other hand, Faster R-CNN shows a slower increase in mAP over each step and fluctuations that do not affect the increasing trend. Although mAP over classes averaged over an IoU of 0.5 for Faster R-CNN is still lower than YOLOv5, at this IoU threshold, it does show better values than at the range from 0.5 to 0.95, which shows that at higher threshold values, it is less accurate.

For Faster R-CNN, the mAP for small, medium, and large objects shows there is better accuracy when the model detects larger objects, as the mAP for small objects does not show convergence and has large fluctuations, especially from 80K to 180K steps, while mAP for medium and large objects does show convergence and less significant fluctuations. This is true for Faster R-CNN as architecture, as it still faces difficulty detecting small objects. Therefore, improvements such as introducing shallow features into the backbone network and ensuring sufficient spatial information for detecting small objects [30] could help improve this value.

It was also observed that Faster R-CNN's average recall with 100 detections per image has fewer fluctuations and is higher as the object size increases. This means that the larger the object, the more accurate the model predicts the object class. Between 10K to 30K steps, the recall value increases significantly and then has small fluctuations after 70K steps. However, Faster R-CNN classification's loss does not show convergence, as YOLOv5's does, but remains less than 0.15 for most of the training. The classification loss for YOLOv5 shows values below 0.04. The loss fluctuates until 120 epochs and then stabilizes approximately to 0.0325. This shows that deviations from predicting "1" for the correct classes and "0" for all the other classes for the object in the box are generally low for both models but rarely happen in YOLOv5's model. This is probably due to the fact that the data set has just two classes.

The Box Classifier localization loss increases over each step (with fluctuations after 80K steps), but overall the values are below 0.1. The loss of RPN localization decreases to about 120K steps, when it increases, and fluctuations become notorious. This value is constantly below 0.110. For YOLOv5, the value initially fluctuates (until about 120 epochs) and then stabilizes to approximately 0.0425. It is mostly below 0.025. Therefore, YOLOv5 is better at having box predictions that cover the objects in the images. While for Faster R-CNN, the objectness loss does not converge, seems to increase, and fluctuates significantly after each step. This means that the model could perform better at box-object IoU prediction. It is also observed that YOLOv5 has an initial spike in this value, which decreases, then slowly increases and fluctuates from 10 to 70 epochs and finally stabilizes at approximately 0.04. For YOLOv5, it is shown that the objectness loss value is mostly below 0.0425, while for Faster R-CNN, it is between 0.25 and 0.29 approximately.

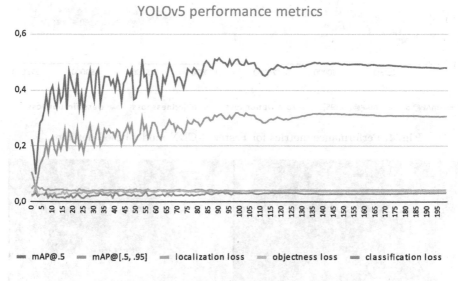

Fig. 3. Performance metrics for YOLOv5 model-based detector.

The Faster R-CNN model had better performance when detecting objects that were alone in the image, as shown in Fig. 5. However, when the image has multiple objects, often overlapping, it usually detects the animals that were less covered by others, as shown in Fig. 6. This suggests that it is challenging for the model to discriminate instance boundaries, given that the features of the instances overlap.

On the other hand, as seen in Fig. 7, YOLOv5 model has better results when detecting objects that were alone in the image, but tends to label images that contain objects as background when the weather and lighting conditions contrast the object, and the background is less visible, as shown in Figs. 8 and 9. This could be improved by adding background images to the dataset, so the model could learn to differentiate it better. Moreover, color augmentations could be tuned to help the model be less sensitive to these changes. YOLOv5 model performs better than Faster R-CNN when dealing with crowded images, inferring several objects in the images, even when there is overlap.

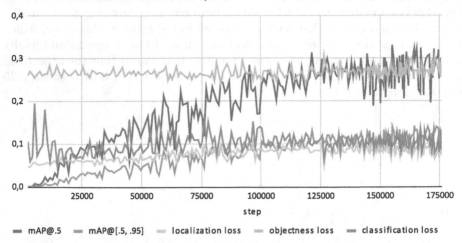

Fig. 4. Performance metrics for Faster R-CNN model-based detector.

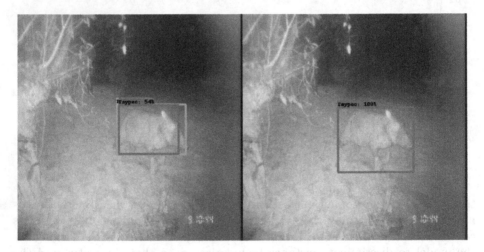

Fig. 5. Example of the detection results for the Faster R-CNN model-based detector when only one object appears in the scene.

These results suggest that, to improve the overall accuracy of Faster R-CNN, it may be helpful to increase the number of steps, as its mAP value increases slowly over each step. Another method would be to use more augmentation options. Also, although Non-Max suppression (NMS) is used, it may be improved by increasing the score threshold to do minimum filtering of proposals. This would stabilize the training and have a better convergence. IoU threshold for NMS could also be reduced, given that this dataset does show a dense distribution of objects in each image, as the animals tend to appear in large numbers

Fig. 6. Example of the detection results for the Faster R-CNN model-based detector when several objects appear in the scene.

Fig. 7. Example of the detection results for the YOLOv5 model-based detector when only one (left) and when several (right) objects appear in the scene.

and close together. Repulsion Loss [33] and AggLoss [37] are two modified losses that can help reduce the sensitivity of results to the NMS IoU threshold, ensuring tighter bounding boxes. Additionally, several attempts have been made to redesign NMS to handle occlusion, such as adaptive NMS [20]. A learning rate schedule could also help improve the overall model accuracy, as it has been done for YOLOv5 using a Once-Cycle Policy and its training stability. The accuracy could also be improved using the alternating optimization algorithm, which allows RPN and Fast R-CNN to be trained to share convolutional features. This

Fig. 8. Example of the detection results for the YOLOv5 model-based detector under different environmental conditions.

is more accurate than approximate joint training, because the latter ignores the weight gradients concerning the region proposals.

Finally, as a preprocessing step, analyzing the geometry of the image objects would also be helpful, as it would allow better tuning of the anchor aspect rations and show the distribution for height-to-width and width-to-height ratios. Moreover, increasing the dataset by adding more images would also benefit both models' performance. This dataset displays a wide intra-class variety with different animal sizes, positions, and angles.

Fig. 9. Example of the detection results for the YOLOv5 model-based detector under different environmental conditions.

3.1 State of Art Based Comparison

The performance of both models was also tested using 300 proposals from the MS COCO dataset [19]. For the validation test, the faster R-CNN-based model achieved a 0.415 mAP@0.5 score and a 0.212 mAP@[0.5, 0.95]. On the other hand, YOLOv5 achieved a 0.54 mAP@0.5 score and a 0.35 mAP@[0.5, 0.95]. As shown in Table 1, for this experiment, Faster R-CNN achieved a mAP@0.5 score of 0.26 and mAP@[0.5, 0.95] of 0.114, while for YOLOv5 the mAP@0.5 score is 0.5525 and mAP@[0.5, 0.95] is 0.37997. Therefore, the YOLOv5 model proved to be more robust, having lower losses and a higher overall mAP value than Faster-RCNN on the MS COCO dataset. In general, Faster R-CNN has shown low performance on small objects compared to other methods [17]. This was seen in this experiment as higher mAP values were achieved for medium and large objects.

Table 1. Metrics Comparison for both models using 300 proposal from the MS COCO dataset.

Model	Average mAP@0.5	Average mAP@[0.5, 0.95]
YOLOv5	0.5525	0.37997
Faster R-CNN	0.26	0.114

In terms of speed (FPS), it is difficult to make comparisons, as they are usually measured at different mAP values. However, a comparison of the accuracy and speed tradeoff may be easier to determine, depending on the application. In terms of Faster R-CNN, the feature extractor used does change the overall mAP achieved in a certain GPU time [17]. In this experiment, Inception V2 was used, which gives one of the lower accuracies compared to other feature extractors [17]. On the other hand, Inception Resnet V2 has been shown to have the highest accuracy at 1 FPS [17]. Therefore, the overall mAP achieved in this experiment by Faster R-CNN could be further improved by using Inception Resnet V2, as it does not require more GPU time to do this.

4 Conclusions and Future Work

In this paper, two state-of-the-art object models were evaluated to detect and classify white-lipped and collared peccary using a dataset of 7,733 camera trap images from the Tiputini Biodiversity Station. YOLOv5 model proved to be more robust, having lower losses and a higher overall mAP value than the Faster-RCNN model, especially for detecting various animals in the scene. Although several optimization techniques were used in each model to improve performance, it is recommended that the dataset is augmented and that the hyperparameters of both models are further tuned. Moreover, it is suggested to train more classes (Tiputini animal species) for a more extensive use of these models. Comparison with other techniques is also required.

Acknowledgment. The authors express their gratitude to the Tiputini Biodiversity Station for providing the data used in this study, which were collected by all researchers and staff working on the TBS-Camara trap project. The station is affiliated with USFQ. Additionally, the authors extend thanks to the Applied Signal Processing and Machine Learning Research Group at USFQ for supplying the computing infrastructure (NVidia DGX workstation) employed for implementing and executing the developed source code.

References

1. Aziz, L., Salam, M.S.B.H., Sheikh, U.U., Ayub, S.: Exploring deep learning-based architecture, strategies, applications and current trends in generic object detection: a comprehensive review. IEEE Access **8**, 170461–170495 (2020)
2. Bass, M.S., et al.: Global conservation significance of ecuador's yasuní national park. PLoS ONE **5**(1), e8767 (2010)
3. Beery, S., Morris, D., Yang, S.: Efficient pipeline for camera trap image review. arXiv preprint arXiv:1907.06772 (2019)
4. Blake, J.G., Mosquera, D., Guerra, J., Loiselle, B.A., Romo, D., Swing, K.: Yasuní-a hotspot for jaguars panthera onca (carnivora: Felidae)? camera-traps and jaguar activity at tiputini biodiversity station, ecuador. Rev. Biol. Trop. **62**(2), 689–698 (2014)
5. Bochkovskiy, A., Wang, C.Y., Liao, H.Y.M.: Yolov4: optimal speed and accuracy of object detection. arXiv preprint arXiv:2004.10934 (2020)
6. Bodla, N., Singh, B., Chellappa, R., Davis, L.: Improving object detection with one line of code. arxiv 2017. arXiv preprint arXiv:1704.04503 (2017)
7. Cao, C., et al.: An improved faster R-CNN for small object detection. IEEE Access **7**, 106838–106846 (2019)
8. Carl, C., Schönfeld, F., Profft, I., Klamm, A., Landgraf, D.: Automated detection of European wild mammal species in camera trap images with an existing and pre-trained computer vision model. Eur. J. Wildl. Res. **66**(4), 1–7 (2020)
9. Cheema, G.S., Anand, S.: Automatic detection and recognition of individuals in patterned species. In: Altun, Y., et al. (eds.) ECML PKDD 2017. LNCS (LNAI), vol. 10536, pp. 27–38. Springer, Cham (2017). https://doi.org/10.1007/978-3-319-71273-4_3
10. Choiński, M., Rogowski, M., Tynecki, P., Kuijper, D.P.J., Churski, M., Bubnicki, J.W.: A first step towards automated species recognition from camera trap images of mammals using AI in a European temperate forest. In: Saeed, K., Dvorský, J. (eds.) CISIM 2021. LNCS, vol. 12883, pp. 299–310. Springer, Cham (2021). https://doi.org/10.1007/978-3-030-84340-3_24
11. Deng, Z., Sun, H., Zhou, S., Zhao, J., Lei, L., Zou, H.: Multi-scale object detection in remote sensing imagery with convolutional neural networks. ISPRS J. Photogramm. Remote. Sens. **145**, 3–22 (2018)
12. Espinosa, S., Celis, G., Branch, L.C.: When roads appear jaguars decline: increased access to an Amazonian wilderness area reduces potential for jaguar conservation. PLoS ONE **13**(1), e0189740 (2018)
13. Girshick, R., Donahue, J., Darrell, T., Malik, J.: Rich feature hierarchies for accurate object detection and semantic segmentation. In: Proceedings of the IEEE Conference on Computer Vision and Pattern Recognition, pp. 580–587 (2014)
14. He, K., Gkioxari, G., Dollár, P., Girshick, R.: Mask R-CNN. In: Proceedings of the IEEE International Conference on Computer Vision, pp. 2961–2969 (2017)
15. He, Y., Weng, Q.: High Spatial Resolution Remote Sensing: Data, Analysis, and Applications. CRC Press, Boca Raton (2018)
16. Ioffe, S., Szegedy, C.: Batch normalization: accelerating deep network training by reducing internal covariate shift. In: International Conference on Machine Learning, pp. 448–456. PMLR (2015)
17. Jocher, G., et al.: ultralytics/yolov5: v6.1 - TensorRT, TensorFlow Edge TPU and OpenVINO Export and Inference (2022). https://doi.org/10.5281/zenodo.6222936

18. Li, Z., Tian, X., Liu, X., Liu, Y., Shi, X.: A two-stage industrial defect detection framework based on improved-yolov5 and optimized-inception-resnetv2 models. Appl. Sci. **12**(2), 834 (2022)
19. Lin, T.-Y., et al.: Microsoft COCO: common objects in context. In: Fleet, D., Pajdla, T., Schiele, B., Tuytelaars, T. (eds.) ECCV 2014. LNCS, vol. 8693, pp. 740–755. Springer, Cham (2014). https://doi.org/10.1007/978-3-319-10602-1_48
20. Liu, S., Huang, D., Wang, Y.: Adaptive NMS: refining pedestrian detection in a crowd. In: Proceedings of the IEEE/CVF Conference on Computer Vision and Pattern Recognition, pp. 6459–6468 (2019)
21. Peng, J., et al.: Wild animal survey using UAS imagery and deep learning: modified faster R-CNN for kiang detection in Tibetan plateau. ISPRS J. Photogramm. Remote. Sens. **169**, 364–376 (2020)
22. Rahman, M.A., Wang, Y.: Optimizing intersection-over-union in deep neural networks for image segmentation. In: Bebis, G., et al. (eds.) ISVC 2016. LNCS, vol. 10072, pp. 234–244. Springer, Cham (2016). https://doi.org/10.1007/978-3-319-50835-1_22
23. Redmon, J., Divvala, S., Girshick, R., Farhadi, A.: You only look once: unified, real-time object detection. In: Proceedings of the IEEE Conference on Computer Vision and Pattern Recognition, pp. 779–788 (2016)
24. Ruder, S.: An overview of gradient descent optimization algorithms. arXiv preprint arXiv:1609.04747 (2016)
25. Norouzzadeh, M.S., et al.: Automatically identifying, counting, and describing wild animals in camera-trap images with deep learning. Proc. Natl. Acad. Sci. **115**(25), E5716–E5725 (2018)
26. Shim, K., Barczak, A., Reyes, N., Ahmed, N.: Small mammals and bird detection using IoT devices. In: 2021 36th International Conference on Image and Vision Computing New Zealand (IVCNZ), pp. 1–6. IEEE (2021)
27. Smith, L.N.: A disciplined approach to neural network hyper-parameters: part 1-learning rate, batch size, momentum, and weight decay. arXiv preprint arXiv:1803.09820 (2018)
28. Srivastava, S., Divekar, A.V., Anilkumar, C., Naik, I., Kulkarni, V., Pattabiraman, V.: Comparative analysis of deep learning image detection algorithms. J. Big Data **8**(1), 1–27 (2021). https://doi.org/10.1186/s40537-021-00434-w
29. Suárez, E., Zapata-Ríos, G., Utreras, V., Strindberg, S., Vargas, J.: Controlling access to oil roads protects forest cover, but not wildlife communities: a case study from the rainforest of Yasuní Biosphere Reserve (Ecuador). Anim. Conserv. **16**(3), 265–274 (2013)
30. Tang, L., Li, F., Lan, R., Luo, X.: A small object detection algorithm based on improved faster RCNN. In: International Symposium on Artificial Intelligence and Robotics 2021, vol. 11884, pp. 653–661. SPIE (2021)
31. Trolliet, F., Vermeulen, C., Huynen, M.C., Hambuckers, A.: Use of camera traps for wildlife studies: a review. Biotechnol. Agron. Soc. Environ. **18**(3), 446–454 (2014)
32. Vargas-Felipe, M., Pellegrin, L., Guevara-Carrizales, A.A., López-Monroy, A.P., Escalante, H.J., Gonzalez-Fraga, J.A.: Desert bighorn sheep (ovis canadensis) recognition from camera traps based on learned features. Eco. Inform. **64**, 101328 (2021)
33. Wang, X., Xiao, T., Jiang, Y., Shao, S., Sun, J., Shen, C.: Repulsion loss: detecting pedestrians in a crowd. In: Proceedings of the IEEE Conference on Computer Vision and Pattern Recognition, pp. 7774–7783 (2018)

34. Weckel, M., Giuliano, W., Silver, S.: Jaguar (panthera onca) feeding ecology: distribution of predator and prey through time and space. J. Zool. **270**(1), 25–30 (2006)
35. Yang, Q., Xiao, D., Lin, S.: Feeding behavior recognition for group-housed pigs with the faster R-CNN. Comput. Electron. Agric. **155**, 453–460 (2018)
36. Yang, Z., Sinnott, R., Ke, Q., Bailey, J.: Individual feral cat identification through deep learning. In: 2021 IEEE/ACM 8th International Conference on Big Data Computing, Applications and Technologies (BDCAT 2021), pp. 101–110 (2021)
37. Zhang, S., Wen, L., Bian, X., Lei, Z., Li, S.Z.: Occlusion-aware R-CNN: detecting pedestrians in a crowd. In: Proceedings of the European Conference on Computer Vision (ECCV), pp. 637–653 (2018)

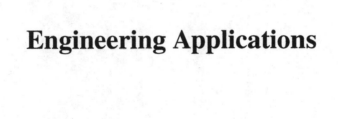

Engineering Applications

Intra-day Electricity Price Forecasting Based on a Time2Vec-LSTM Model

Sergio Cantillo-Luna[1] , Ricardo Moreno-Chuquen[2]([⊠]) ,
and Jesus Lopez-Sotelo[1]

[1] Universidad Autónoma de Occidente, Cali 760030, Colombia
{sergio.cantillo,jalopez}@uao.edu.co
[2] Universidad Icesi, Cali, Colombia
rmoreno@icesi.edu.co

Abstract. This paper presents a novel deep neural network architecture, combining stacked LSTM and Time2Vec layers, to predict electricity prices up to eight hours ahead-an essential input for future decision-making tools. We assess this model using hourly wholesale electricity price data from Colombia, comparing it to state-of-the-art time series and machine learning forecasting methods, including SARIMA, Holt-Winters, XGBoost, and MLP. The results demonstrate that our model excels in accurately modeling non-linearity and explicitly characterizing data behavior, yielding more precise price predictions. Particularly, the Time2Vec layer significantly aids in capturing temporal relationships between input and output data. This framework shows promise in enhancing the precision of electric price forecasts, offering valuable insights for the energy sector's decision-making.

Keywords: Decision-making · Electricity Price Forecasting (EPF) · LSTM · Time2Vec

1 Introduction

The integration of Distributed Energy Resources (DER) into existing power systems has gained increasing significance due to the growing demand for sustainable and reliable energy sources. DERs include residential-scale photovoltaic (PV) systems, energy storage systems (ESS), electric vehicles (EV), and demand response programs (DR), offering technical and environmental advantages [8,35]. However, the integration of DERs presents operational, reliability, and safety challenges for power grids [7]. One critical issue is the unpredictability and variability of renewable energy sources, leading to abrupt fluctuations in electricity supply and load.

These circumstances make it challenging to balance energy supply and demand in real-time, considering various constraints (e.g., economic, operational, social welfare). Consequently, accurately determining their price becomes intricate, with unforeseen and sudden price spikes becoming more frequent, further complicating their temporal behavior [32].

A. D. Orjuela-Cañón et al. (Eds.): ColCACI 2023, CCIS 1865, pp. 107–121, 2024.
https://doi.org/10.1007/978-3-031-48415-5_8

Accurate electricity price forecasting several hours ahead is essential for decision-making by various stakeholders, including DER owners, power system operators (DSOs), and emerging business models [13]. This is necessary to effectively manage these situations and make better-informed decisions considering the ongoing development and evolution of power grids. Electricity prices also play a crucial role in resource allocation, ensuring overall reliability and efficiency [43]. Furthermore, these prices directly influence the behavior of all market players according to their interests.

To address these challenges, this paper focuses on intra-day electricity price forecasting using a proposed Time2Vec layered long short-term memory (LSTM) architecture, referred to as T2V-SLSTM. This model features a hybrid architecture that combines stacked LSTM neural networks with an explicit time variable embedding referred to as Time2Vec [17]. For this study, we consider only historical hourly electricity prices in the Colombian wholesale market to assess the performance of the proposed model.

In addition, this paper conducts a comparative analysis of the proposed model's performance with task-related time series statistical analysis (TSA) forecasting models such as Holt-Winters and seasonal ARIMA (SARIMA), as well as machine and deep learning models such as Extreme Gradient Boosting (XGB) and Multi-Layer Perceptron (MLP), respectively. The objective of this comparison is to provide insights into new decision-making tools for DERs. This analysis uses performance metrics such as RMSE, MAE, and MAPE.

The structure of this paper unfolds as follows: Sect. 1 presents the introduction, while Sect. 2 details the methodology and the proposed T2V-LSTM model, along with a brief description of considered baseline models. Subsequently, Sect. 3 offers some experimental results and baseline comparisons with the corresponding analysis. Finally, Sect. 4 summarizes the paper, highlights key findings, and provides concluding remarks.

1.1   Literature Review

In electricity markets, trading takes place in a dynamic and unpredictable environment with technical, operational, and economic constraints [29]. Accurate electricity price forecasting is crucial in this context, impacting decision-makers, market players, and power producers alike [34].

Electricity price forecasting is essential for adjusting production plans, making real-time decisions, and optimizing resource allocation while maximizing economic benefits and minimizing risks. It also contributes to power grid stability, even in scenarios with new energy resources like DERs, despite the challenges posed by seasonal fluctuations and highly non-linear, time-varying features [21].

To address these challenges, researchers and decision-makers employ various approaches, categorized into five primary domains: fundamental methods, multi-agent methods, reduced-form methods, statistical models (including time-series analysis methods), and computational intelligence models [15]. This study primarily focuses on statistical and computational intelligence models [40].

Fundamental methods use explicit formulas incorporating generation units and production costs, but they require extensive data for accurate modeling [22]. Multi-agent models simulate system operation through diverse agents but are mainly used for qualitative purposes [38]. Reduced-form models describe statistical properties of electricity prices for risk management, limiting their ability to predict hourly prices [41].

Statistical and computational intelligence models have become mature in electricity price forecasting due to their accuracy in short-term predictions without detailed system modeling [6]. Initially, statistical time series analysis forecasting models such as linear regression, moving averages, and auto-regressive models were used alongside advanced techniques like ARIMA-based approaches, exponential smoothing, Box-Jenkins, state-space modeling, and hybrid statistical models [14]. These models utilize historical data and exogenous variables for forecasting.

Several studies, including [1, 27, 31], have explored electricity price forecasting using these models. For example, [31] integrated an ARIMA model with a neural network, improving forecast accuracy. Similarly, [27] developed ARIMA, SARIMA, GARCH, and hybrid models, demonstrating their competitiveness. Lastly, in [1], SARIMAX with exogenous variables outperformed other models (SARIMA, SARIMAX and ARIMA) in predicting day-ahead electricity prices in Germany. However, all the statistical time series forecasting approaches presented use linear forecasts, which limits its accuracy with data with high variability and volatility, especially for predicting multiple future values. They excel when data seasonality is low but may not be ideal for all scenarios.

Machine and deep learning models have shown promise in electricity price forecasting, including support vector machines (SVM) [3, 30], k-nearest neighbors [5, 16] (KNN), tree-based models [2, 25], convolutional neural nets (CNN) [18, 20], multi-layer perceptrons (MLP) [19, 36] including extreme machine learning models (ELM) [10], and recurrent networks such as LSTM and GRU-based models [23, 43], as well as ensembles [4, 42]. These approaches offer potential improvements in forecasting accuracy for data with high variability and complex patterns.

2   Methodology

The methodology in this paper includes three phases. The first phase (1) refers to data collection and processing, which includes converting the electricity spot price time series into a compatible dataset for the proposed Time2Vec-LSTM based model. This dataset is then divided into training, validation, and test sets. The second phase (2) focuses on developing and tuning the hyperparameters of the proposed model using the training and validation datasets obtained from the previous stage. Finally, the proposed model is assessed (3) against different forecasting models, including SARIMA, XGBoost, Multi-layer perceptron and Holt-Winters models using various forecasting performance metrics. Figure 1 provides an overview of the methodology used for this study.

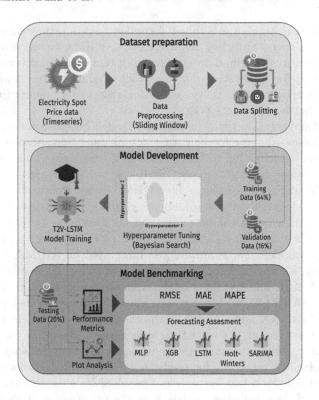

Fig. 1. Framework of the proposed methodology.

2.1 Data Collection and Processing

The time series data, such as electricity prices in Colombia, often exhibit varying correlations across different time scales, such as daily, weekly, or monthly patterns. Deep learning models, including our proposed model, face challenges when dealing with regression tasks involving this type of data, notably multicollinearity issues that can hinder accurate interpretation of variable dependencies across various time intervals.

To effectively adapt our proposed model, we transform the one-dimensional time series data into a dataset structured as [samples, time lags, features] along with output variables. This format aligns seamlessly with LSTM-based learning tasks. To integrate data input into deep learning methods, we employ the sliding window technique, which facilitates the training and forecasting processes.

In our study, we use a prediction window spanning 48 h (time lags) to represent the hourly electricity price data from the past two days. We also establish an 8-hour forecasting horizon. These choices result from a comprehensive analysis of time series data and model tuning experiments. They are based on the recognition of multi-day patterns corresponding to daily price cycles, striking a balance between incorporating relevant historical data and minimizing computational complexity, while considering precision and scalability.

The dataset utilized in this study comprises wholesale spot price data of electricity in Colombia, spanning from January 1, 2018, to December 31, 2022, with hourly frequency. The data was acquired from the web platform of the market operator, Sinergox from XM [39].

2.2   Proposed Model

This section introduces and discusses the development of the proposed hybrid architecture for the forecasting model as shown in Fig. 2. The T2V-SLSTM architecture combines Time2Vec embedding [17], known for its temporal representation capabilities, with stacked LSTM models, recognized for robust learning capabilities. This combined approach enhances the dependability and precision of electricity price forecasts.

The forecasting model follows this structure: it starts with a Time2Vec (T2V) embedding as the input layer, followed by a stacked Long Short-Term Memory (LSTM) architecture consisting of three layers. The choice of three layers resulted from tests balancing accuracy and computational cost. Next, the model includes a dense output layer with a number of units corresponding to an 8-hour prediction horizon, aligning with the intraday market's three 8-hour blocks in Colombia [11]. This choice is crucial, as multi-step ahead predictions are more beneficial for stakeholders than single-step predictions [33].

The preprocessed electricity price data is fed into the initial Time2Vec layer, which generates a unique and embedded representation of time. Importantly, this representation is model-agnostic and can integrate seamlessly into various neural network architectures. Time2Vec simplifies the feature engineering process, enhancing the model's overall modeling capacity. This approach aligns with the deployment methodology detailed in Eq. (1).

$$T2V(\tau)[i] = \begin{cases} \omega_i \cdot \tau + \phi_i & \text{if } i = 0, \\ F\left(\omega_i \cdot \tau + \phi_i\right), & \text{if } 1 \leq i \leq k \end{cases} \tag{1}$$

Within the T2V layer, referred to as T2V, the raw data (τ) undergoes processing through adjustable parameters $(\omega$ and $\phi)$ as part of the model learning process. The activation function (F) employed in this layer captures periodic patterns within the data using a sine wave function. The hyperparameter (k) determines the output dimensionality of this layer, influencing the number of sinusoidal patterns learned. Ultimately, the T2V layer generates a composite representation consisting of acquired sinusoidal patterns along with a linear representation of the input at $i = 0$.

The processed data then flows into a series of stacked LSTM layers, chosen for their ability to analyze short and long-term temporal patterns accurately. To mitigate overfitting, a dropout process of 10% is implemented between each LSTM layer.

Within this model, each LSTM layer can selectively retain or discard information through three gate structures: the input gate (I_t), the forget gate (F_t),

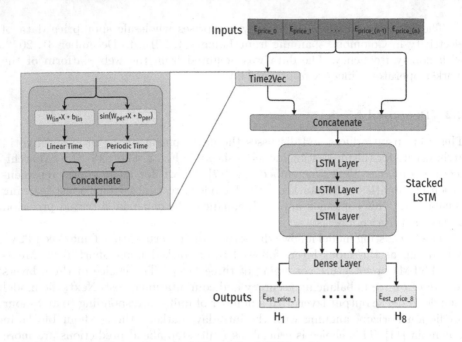

Fig. 2. Proposed Time2Vec-LSTM architecture.

and the output gate (O_t) [24]. The LSTM cell operation is defined in Eqs. (2) to (6), where "·" represents an element-wise product.

$$F_t = \sigma(W_f \cdot X_t + W_{hf} \cdot H_{t-1} + b_f) \tag{2}$$

$$I_t = \sigma(W_i \cdot X_t + W_{hi} \cdot H_{t-1} + b_i) \tag{3}$$

$$O_t = \sigma(W_o \cdot X_t + W_{ho} \cdot H_{t-1} + b_o) \tag{4}$$

$$C_t = F_t \cdot C_{t-1} + I_t \cdot \tanh(W_c \cdot X_t + W_{hc} \cdot H_{t-1} + b_c) \tag{5}$$

$$H_t = O_t \cdot \tanh(C_t) \tag{6}$$

We introduce weight matrices denoted as W_i, W_f, W_o, W_c, W_{hi}, W_{hf}, W_{ho}, and W_{hc}, associated with various gates, and biases represented by b_i, b_f, b_o, and b_c. C_t and C_{t-1} refer to the current and previous states of the cell outputs, while H_t and H_{t-1} indicate the current and preceding hidden states.

Following this, the stacked LSTM layers generate an output, which is then fed into the dense output layer with a linear activation function. The model produces output values within the $(-1, 1)$ range, which are subsequently rescaled to their original range. To evaluate the model's performance, performance metrics as described in Sect. 2.4 are utilized.

2.3 Hyperparameter Tuning

To optimize the proposed model's performance, we carefully fine-tuned hyperparameters to minimize the mean squared error (MSE) loss, which quantifies the difference between actual ($E_p^{(i)}$) and predicted ($\hat{E}_p^{(i)}$) electricity prices, as defined in Eq. (7). We utilized the Adam algorithm for optimization and implemented early stopping and dynamic learning rate reduction during model training and tuning, ranging from 0.001 to 0.00001, to prevent overfitting and improve forecasting accuracy.

$$\text{MSE} = \frac{1}{N} \sum_{i=1}^{N} (E_p^{(i)} - \hat{E}_p^{(i)})^2 \tag{7}$$

For Bayesian hyperparameter search, we explored a flexible yet constrained range for the model's layer units, excluding the last dense layer. These parameters included fixed values and stochastic uniform (U) variations. To ensure fair model comparisons, our search involved 200 training epochs, employed the *Tanh* activation function for the LSTM layer, and used batch sizes of 128 samples. This hyperparameter search was facilitated by the 'hypertune-keras' package, which builds upon hyperopt [9]. The search space and tuning results are summarized in Table 1. Consequently, this model architecture was trained with a total of 245,032 parameters.

Table 1. Hyperparameter search space and tuning values for the proposed model

Model Layer	Hyperparameter search space	
	Value range	Selected value
Time2Vec Layer	64 + U(0,64)	105
LSTM Layer 1	64 + U(0,64)	108
LSTM Layer 2	64 + U(0,64)	104
LSTM Layer 3	64 + U(0,64)	79

2.4 Performance Metrics

The assessment of the proposed forecasting model involved utilizing standard time series forecasting performance metrics, which encompass the root mean square error (RMSE), mean absolute error (MAE), and mean absolute percentage error (MAPE). The selected performance metrics for this investigation are introduced in Eqs. (8) to (10).

In these equations, E_p represents the actual electricity price value, while $\hat{E}_p$ and $\bar{E}_p$ correspond to the electricity price forecasted by the model and its mean, respectively. Dimensionless metrics and units were deliberately chosen to facilitate comparative analysis with other forecasting models and to demonstrate the influence of these predictions on decision-making tools for managing distributed energy resources (DER). This entails evaluating economic risk using MAE as a risk-neutral metric and RMSE as a risk-averse metric.

$$\text{RMSE} = \sqrt{\frac{1}{N} \sum_{i=1}^{N} (E_p^{(i)} - \hat{E_p}^{(i)})^2} \tag{8}$$

$$\text{MAE} = \frac{1}{N} \sum_{i=1}^{N} \left| E_p^{(i)} - \hat{E_p}^{(i)} \right| \tag{9}$$

$$\text{MAPE} = \frac{1}{N} \sum_{i=1}^{N} \frac{\left| E_p^{(i)} - \hat{E_p}^{(i)} \right|}{E_p^{(i)}} * 100\% \tag{10}$$

3 Experimental Results and Analysis

This experiment employed an hourly electricity price dataset from Colombia. To assess our model's performance, we compared it to established baseline models frequently used in energy-related time series forecasting, including Seasonal ARIMA (SARIMA), Holt-Winters, Extreme Gradient Boosting (XGB), and Multi-Layer Perceptron models.

All data processing and model development were carried out on a personal computer (PC) running the Windows® operating system, equipped with an Intel® Core i5+ 10300H processor operating at 2.5 GHz and 16.00 GB of RAM. For hyperparameter tuning, model and baseline training, and benchmarking, we utilized the Google® Colab platform, implementing it with the Keras [12], Scikit-learn [26], and Statsmodels [28] Python APIs.

3.1 Exploratory Data Analysis

This dataset included 43,824 samples from 2018 to 2022, reflecting stepwise DER penetration in the Colombian wholesale electricity market. Data was divided into two subsets: January 2018 to December 2021 (35,009 samples, 80% of the data) for training and validation, and January 2022 to December 2022 (8,705 samples, 20% of the data) for testing.

An initial hourly time series analysis (see Fig. 3) confirms a strong connection between electricity demand fluctuations and electricity price variations, consistent with prior research [8,37]. Peak hours (19 and 20) correlate with higher prices due to increased power consumption, while off-peak hours (3 and 4) show the opposite trend. Rare instances of exceptionally high prices (over 650 COP/kWh), attributed to climatic or operational issues, constitute less than 2.5% of the dataset. To mitigate their impact on the forecasting model, we applied a natural logarithm transformation, as shown in Eq. (11).

$$E_{p,t_{trans}} = \ln(E_{p,t} + 1) \tag{11}$$

$E_{p,t}$ denotes electricity spot price at hour t, and $E_{p,t_{trans}}$ is the transformed data. To ensure model compatibility, we standardized the data within $(-1, 1)$. When making predictions, we reverse this process to re-scale the data.

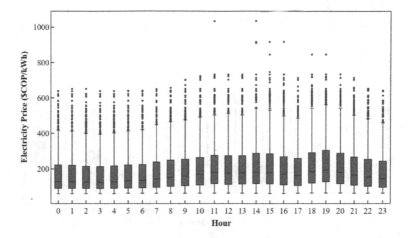

Fig. 3. Temporal Analysis of Electricity Prices: Box Plot Representation by Hour.

This hourly pattern holds on a month-by-month basis as depicted in Fig. 4. Quarterly analysis reveals Q2 and Q4 as having the most price outliers, with June having the highest frequency (month with low variability). In contrast, February, October, and December exhibit the greatest price variability due to seasonal demand fluctuations, weather conditions, and economic activity.

Likewise, Table 2 provides statistical data on electricity prices, organized by subset and quarter. In 2022, notable price fluctuations, especially in Q2 (ranging from \$270 to \$1035 COP/kWh in one hour), make it a focal point for our research. We use this dataset to evaluate our T2V-LSTM model against baseline models considering operational scenarios variability.

Based on the 2022 price data behavior, we have defined operational scenarios with low (Q1), moderate (Q3), high (Q4), and very high (Q2) variability for testing purposes. These scenarios, determined by factors like Interquartile Range (IQR) and outliers frequency, will aid in evaluating the performance of the model prediction.

3.2   Model Performance Assessment

To assess intra-day electricity price forecasting performance, we compared the T2V-LSTM model with baseline methods across four quarterly scenarios in 2022: Q1 (Jan–Mar), Q2 (Apr–Jun), Q3 (Jul–Sep), and Q4 (Oct–Dec). Statistical time series models $SARIMA(2, 0, 1)_{24}$ and Holt-Winters were used as baselines. SARIMA parameters (p, q, and d) were determined from autocorrelation and partial autocorrelation plots, while Holt-Winters parameters, including α, were optimized to minimize mean squared error (MSE). Note that this process used log-transformed time series data without windowing.

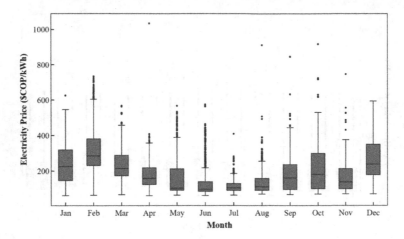

Fig. 4. Temporal Analysis of Hourly Electricity Prices: Box Plot Representation by Month.

Table 2. Descriptive statistics for Colombian wholesale market electricity prices.

Subset	Quarter	Mean	Max.	Min.	Std.	Skew.	Kurt.
2018–2021	Q1	249.20	646.02	61.76	105.20	0.71	0.17
	Q2	167.87	573.19	61.46	102.66	1.26	1.01
	Q3	132.26	910.47	64.42	64.81	2.02	6.49
	Q4	197.57	745.50	69.86	112.88	1.24	1.18
2022	Q1	302.23	731.24	93.49	134.44	1.52	1.60
	Q2	117.68	1035.13	89.06	45.57	9.06	153.72
	Q3	169.57	845.12	97.10	90.88	1.89	5.71
	Q4	274.46	917.14	109.35	115.78	0.89	1.61

Machine learning (ML) and deep learning (DL) models like XGBoost (XGB) and Multi-Layer Perceptron (MLP) respectively were also baselines. Hyperparameter tuning, including Bayesian search as in the T2V-LSTM, was meticulously performed for each model. This tuning process required forecasting windows (sliding window method).

This assessment covered various operational scenarios with low to moderate variability (Q1 and Q3) where all models performed similarly. Our T2V-SLSTM model showed slightly better results with around a 1.3% MAE reduction and 4.1% RMSE reduction compared to the best baseline (MLP). However, significant differences emerged in scenarios with higher variability and fast changing patterns (Q4 and especially Q2). Our model outperformed baseline models with approximately 7.7% MAE reduction and 3.8% RMSE reduction in high variability cases. This adaptability (i.e., extract meaningful representations and capture the intricate temporal dependencies) is particularly evident in Q4. This scenario presents a significant challenge for stakeholders for decision-making (see Fig. 5).

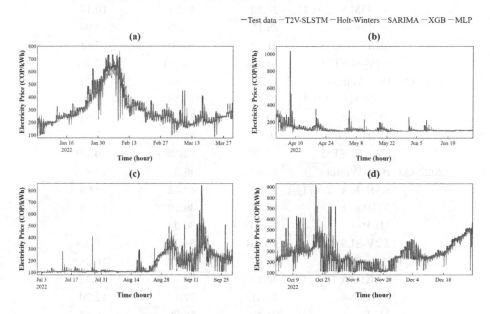

Fig. 5. Comparative Analysis of Electricity Price Forecasting: T2V-SLSTM Model vs. Baseline Models Across Quarters in 2022 (a) First Quarter, (b) Second Quarter, (c) Third Quarter, (d) Fourth Quarter.

The T2V-SLSTM model excels in modeling hourly spot electricity prices, surpassing statistical time series models, especially in long-term patterns and high, consistent hourly variability across all quarters. Log transformation significantly improved overall performance by approximately 8.8%. Furthermore, statistical models struggled with hourly fluctuations, with MAPE increasing by more than 3% when using all data. In contrast, the T2V-SLSTM model demonstrated an overall 4–7% reduction in average electricity price errors compared to these models.

When comparing the computational intelligence benchmark models XGB and MLP with the proposed method, the differences are not significant, with an overall forecast error performance gap of less than 15%. Nonetheless, the proposed model still outperforms them, leading to an overall reduction in error. This, in turn, can increase the level of confidence in decision making among various stakeholders (see Table 3).

Table 3. Performance benchmarking

Data	Model	Performance metrics		
		MAE ($\frac{\$COP}{kWh}$)	RMSE ($\frac{\$COP}{kWh}$)	MAPE (%)
2022-Q1	Holt-Winters	27.43	49.30	9.97
	SARIMA $(2,0,1)_{24}$	28.32	49.28	10.17
	XGBoost	28.15	45.89	8.96
	MLP	21.89	36.70	7.81
	T2V-SLSTM	**21.68**	**35.20**	**7.34**
2022-Q2	Holt-Winters	11.74	40.86	7.98
	SARIMA $(2,0,1)_{24}$	12.12	40.58	8.28
	XGBoost	10.84	31.74	7.64
	MLP	10.86	34.11	7.52
	T2V-SLSTM	**9.25**	**31.28**	**6.30**
2022-Q3	Holt-Winters	21.89	49.37	10.85
	SARIMA $(2,0,1)_{24}$	22.28	49.52	11.06
	XGBoost	19.47	39.19	10.08
	MLP	17.50	36.09	8.93
	T2V-SLSTM	**16.16**	**34.72**	**8.18**
2022-Q4	Holt-Winters	39.07	74.21	16.68
	SARIMA $(2,0,1)_{24}$	40.18	73.59	16.42
	XGBoost	31.51	57.47	12.70
	MLP	31.47	56.08	13.45
	T2V-SLSTM	**28.39**	**53.05**	**11.70**
All data	Holt-Winters	25.03	54.85	11.39
	SARIMA $(2,0,1)_{24}$	25.72	54.61	11.51
	XGBoost	22.62	45.02	9.87
	MLP	20.36	41.63	9.37
	T2V-SLSTM	**18.86**	**39.54**	**8.40**

4   Conclusions

This paper introduces the Time2Vec - Stacked LSTM (T2V-SLSTM) hybrid forecasting model for multi-step forecasting of hourly electricity prices in the Colombian wholesale market. T2V-SLSTM is tailored to address the challenges posed by the integration of distributed energy resources (DERs) and the dynamic nature of electricity prices in evolving market scenarios.

T2V-SLSTM improves hourly electricity price forecasts with an 8-h lead time considering policy perspectives, outperforming traditional statistical and computational intelligence forecasting models. This enhanced performance is attributed to LSTM's capacity to capture short and long-term patterns in price behavior, augmented by the Time2Vec layer representation of temporal and data features.

Reduced prediction errors (RMSE and MAE) highlight the model's suitability for decision-making tools catering to various risk preferences.

In summary, T2V-SLSTM offers a promising framework for enhancing decision-making in the electricity market, considering diverse variability scenarios. It empowers stakeholders to optimize their energy trading activities even in dynamic markets.

Future research will explore the impact of exogenous variables like hourly load on model performance. Additionally, more robust probabilistic forecasting architectures will be employed to account for substantial data variability, making this approach better suited for scenario generation in decision-making tools.

Acknowledgements. The authors would like to thank the support of the Universidad Icesi and Universidad Autónoma de Occidente in Cali, Colombia.

References

1. Abunofal, M., Poshiya, N., Qussous, R., Weidlich, A.: Comparative analysis of electricity market prices based on different forecasting methods. In: 2021 IEEE Madrid PowerTech. IEEE, June 2021. https://doi.org/10.1109/powertech46648.2021.9495034
2. Albahli, S., Shiraz, M., Ayub, N.: Electricity price forecasting for cloud computing using an enhanced machine learning model. IEEE Access **8**, 200971–200981 (2020)
3. Ali, M., Khan, Z.A., Mujeeb, S., Abbas, S., Javaid, N.: Short-term electricity price and load forecasting using enhanced support vector machine and k-nearest neighbor. In: 2019 Sixth HCT Information Technology Trends (ITT), pp. 79–83. IEEE (2019)
4. Alkawaz, A.N., Abdellatif, A., Kanesan, J., Khairuddin, A.S.M., Gheni, H.M.: Day-ahead electricity price forecasting based on hybrid regression model. IEEE Access **10**, 108021–108033 (2022)
5. Ashfaq, T., Javaid, N.: Short-term electricity load and price forecasting using enhanced KNN. In: 2019 International Conference on Frontiers of Information Technology (FIT). IEEE, December 2019. https://doi.org/10.1109/fit47737.2019.00057
6. Brusaferri, A., Matteucci, M., Portolani, P., Vitali, A.: Bayesian deep learning based method for probabilistic forecast of day-ahead electricity prices. Appl. Energy **250**, 1158–1175 (2019)
7. Cantillo-Luna, S., Moreno-Chuquen, R., Chamorro, H.R., Sood, V.K., Badsha, S., Konstantinou, C.: Blockchain for distributed energy resources management and integration. IEEE Access **10**, 68598–68617 (2022)
8. Cantillo-Luna, S., Moreno-Chuquen, R., Lopez-Sotelo, J.A.: Intra-day electricity price forecasting based on a Time2Vec-LSTM neural network model. In: 2023 IEEE Colombian Conference on Applications of Computational Intelligence (ColCACI), pp. 1–6. IEEE (2023)
9. Cerliani, M.: Keras-hypetune (2023). https://github.com/cerlymarco/keras-hypetune
10. Chen, X., Dong, Z.Y., Meng, K., Xu, Y., Wong, K.P., Ngan, H.: Electricity price forecasting with extreme learning machine and bootstrapping. IEEE Trans. Power Syst. **27**(4), 2055–2062 (2012)

11. Comisión de Regulación de Energía y Gas: Documento CREG-114 - Modernización de mercado de energía mayorista (Despacho vinculante, mercados intradiarios y servicios complementarios), September 2021

12. Gulli, A., Pal, S.: Deep Learning with Keras. Packt Publishing Ltd., Birmingham (2017)

13. Heidarpanah, M., Hooshyaripor, F., Fazeli, M.: Daily electricity price forecasting using artificial intelligence models in the Iranian electricity market. Energy **263**, 126011 (2023)

14. Hyndman, R.J., Athanasopoulos, G.: Forecasting: Principles and Practice. OTexts, Heathmont (2018)

15. Jiang, P., Nie, Y., Wang, J., Huang, X.: Multivariable short-term electricity price forecasting using artificial intelligence and multi-input multi-output scheme. Energy Econ. **117**, 106471 (2023)

16. Johannesen, N.J., Kolhe, M., Goodwin, M.: Deregulated electric energy price forecasting in NordPool market using regression techniques. In: 2019 IEEE Sustainable Power and Energy Conference (iSPEC). IEEE, November 2019. https://doi.org/10.1109/ispec48194.2019.8975173

17. Kazemi, S.M., et al.: Time2vec: learning a vector representation of time. arXiv preprint arXiv:1907.05321 (2019)

18. Khan, Z.A., et al.: Short term electricity price forecasting through convolutional neural network (CNN). In: Barolli, L., Amato, F., Moscato, F., Enokido, T., Takizawa, M. (eds.) WAINA 2020. AISC, vol. 1150, pp. 1181–1188. Springer, Cham (2020). https://doi.org/10.1007/978-3-030-44038-1_108

19. Lago, J., De Ridder, F., Vrancx, P., De Schutter, B.: Forecasting day-ahead electricity prices in Europe: the importance of considering market integration. Appl. Energy **211**, 890–903 (2018)

20. Li, W., Becker, D.M.: Day-ahead electricity price prediction applying hybrid models of LSTM-based deep learning methods and feature selection algorithms under consideration of market coupling. Energy **237**, 121543 (2021)

21. de Marcos, R.A., Bello, A., Reneses, J.: Short-term forecasting of electricity prices with a computationally efficient hybrid approach. In: 2017 14th International Conference on the European Energy Market (EEM), pp. 1–6. IEEE (2017)

22. de Marcos, R.A., Bello, A., Reneses, J.: Electricity price forecasting in the short term hybridising fundamental and econometric modelling. Electric Power Syst. Res. **167**, 240–251 (2019)

23. Memarzadeh, G., Keynia, F.: Short-term electricity load and price forecasting by a new optimal LSTM-NN based prediction algorithm. Electric Power Syst. Res. **192**, 106995 (2021)

24. Mittal, D.A., Liu, S., Xu, G.: Electricity price forecasting using convolution and LSTM models. In: 2020 7th International Conference on Behavioural and Social Computing (BESC), pp. 1–4. IEEE (2020)

25. Orenc, S., Acar, E., Ozerdem, M.S.: The electricity price prediction of Victoria city based on various regression algorithms. In: 2022 Global Energy Conference (GEC). IEEE, October 2022. https://doi.org/10.1109/gec55014.2022.9986605

26. Pedregosa, F., et al.: Scikit-learn: machine learning in Python. J. Mach. Learn. Res. **12**, 2825–2830 (2011)

27. Rajan, P., Chandrakala, K.V.: Statistical model approach of electricity price forecasting for Indian electricity market. In: 2021 IEEE Madras Section Conference (MASCON). IEEE, August 2021. https://doi.org/10.1109/mascon51689.2021.9563474

28. Seabold, S., Perktold, J.: Statsmodels: econometric and statistical modeling with python. In: 9th Python in Science Conference (2010)
29. Shao, Z., Zheng, Q., Liu, C., Gao, S., Wang, G., Chu, Y.: A feature extraction-and ranking-based framework for electricity spot price forecasting using a hybrid deep neural network. Electric Power Syst. Res. **200**, 107453 (2021)
30. Shrivastava, N.A., Khosravi, A., Panigrahi, B.K.: Prediction interval estimation of electricity prices using PSO-tuned support vector machines. IEEE Trans. Industr. Inf. **11**(2), 322–331 (2015)
31. Skopal, R.: Short-term hourly price forward curve prediction using neural network and hybrid ARIMA-NN model. In: 2015 International Conference on Information and Digital Technologies. IEEE, July 2015. https://doi.org/10.1109/dt.2015. 7222993
32. Sridharan, V., Tuo, M., Li, X.: Wholesale electricity price forecasting using integrated long-term recurrent convolutional network model. Energies **15**(20), 7606 (2022)
33. Su, H., Peng, X., Liu, H., Quan, H., Wu, K., Chen, Z.: Multi-step-ahead electricity price forecasting based on temporal graph convolutional network. Mathematics **10**(14) (2022). https://doi.org/10.3390/math10142366
34. Tan, Y.Q., Shen, Y.X., Yu, X.Y., Lu, X.: Day-ahead electricity price forecasting employing a novel hybrid frame of deep learning methods: a case study in NSW, Australia. Electric Power Syst. Res. **220**, 109300 (2023). https://doi.org/10.1016/ j.epsr.2023.109300
35. Trivedi, R., et al.: Community-based microgrids: literature review and pathways to decarbonise the local electricity network. Energies **15**(3), 918 (2022)
36. Udaiyakumar, S., Chinnadurrai, C., Anandhakumar, C., Ravindran, S.: Electricity price forecasting using multilayer perceptron optimized by particle swarm optimization. In: 2022 Smart Technologies, Communication and Robotics (STCR), pp. 1–6. IEEE (2022)
37. Urbano Buriticá, S.N., González Pérez, L.F., et al.: Proyección de corto plazo para el precio de bolsa de energía en el mercado colombiano (2022)
38. Weron, R.: Electricity price forecasting: a review of the state-of-the-art with a look into the future. Int. J. Forecast. **30**(4), 1030–1081 (2014)
39. XM Colombia: Portal BI de variables del mercado eléctrico colombiano SINER-GOX. https://sinergox.xm.com.co/trpr/Paginas/Historicos/Historicos.aspx
40. Yang, W., Wang, J., Niu, T., Du, P.: A hybrid forecasting system based on a dual decomposition strategy and multi-objective optimization for electricity price forecasting. Appl. Energy **235**, 1205–1225 (2019)
41. Zhang, F., Fleyeh, H.: A review of single artificial neural network models for electricity spot price forecasting. In: 2019 16th International Conference on the European Energy Market (EEM), pp. 1–6. IEEE (2019)
42. Zhang, F., Fleyeh, H., Bales, C.: A hybrid model based on bidirectional long short-term memory neural network and Catboost for short-term electricity spot price forecasting. J. Oper. Res. Soc. **73**(2), 301–325 (2022)
43. Zhou, S., Zhou, L., Mao, M., Tai, H.M., Wan, Y.: An optimized heterogeneous structure LSTM network for electricity price forecasting. IEEE Access **7**, 108161–108173 (2019)

Radio Frequency Pattern Matching - Subscriber Location in 5G Networks

René Játiva E.$^{(\boxtimes)}$ (ID), Oliver Caisaluisa (ID), Katty Beltrán (ID),
and Martín Gavilánez (ID)

Colegio de Ciencias e Ingeniería, Universidad San Francisco de Quito, Quito, Ecuador
rjativa@usfq.edu.ec,
{oliver.caisaluisa,katty.beltran,martin.gavilanez}@ieee.org

Abstract. Received Signal Strength measures have been collected at the Base Station antenna array of a wireless network operating at 28 GHz mmWaves, and virtually deployed using Open Street Maps and Matlab®. These radio frequency patterns imprinted by a geolocated subscriber transmitting along the campus, have been used to automatically discover the characteristics of the area of interest by using k-means clustering into the proposed unsupervised method. This technique has been integrated into supervised ML methods based on K-Nearest Neighbors, in order to provide an accurate estimation of the subscriber position by performing the match between the received RF patterns and the stored fingerprints. New results exhibit an improved accuracy over previous works based on supervised ML methods. Furthermore, the impact of the operation frequency band over positioning has been evaluated within the range of 3.7–30 GHz. Results show that accuracy degrades at lower frequencies and some mitigation methods are discussed.

Keywords: 5G wireless networks · mmWave · subscriber location · fingerprinting · Radio Frequency Pattern Matching · RFPM · K-Means Clustering · k-means

1 Introduction

Current 5G mobile communications systems are almost ubiquitous, exhibiting unlimited coverage anywhere and anytime for any device. They are capable of providing access to a very large number of devices and also challenging services related to the Internet of Things (IoT). An important part of 5G technology lies in its wide spectral range, defined by ITU-R as Frequency Range 1 (FR1) for operation frequencies up to 7125 MHz, and Frequency Range 2 (FR2) for frequencies up to 52600 MHz. [9]. FR1 is designed to carry most of the traditional cellular traffic, whilst FR2 is designed to provide short distance and very high data rates. Another important piece of this new range of services involves the location of devices accessing the network [15]. Furthermore, 5G has introduced

This work has been partially funded with USFQ Poli-Grant 18012.

new features, such as ultradense networks, large-scale antenna arrays, and beam-forming, providing high capacity, high reliability, and low delays in broad cover-age areas. All these new features are suitable for performing and improving wire-less positioning [18]. Not only traditional geometrical methods based on network and User Equipment (UE) measures, such as Time Of Arrival (TOA), Time Trip Time (TTT), Time Differences Of Arrival (TDOA), Observed TDOA (OTDOA), Angle Of Arrival (AOA), etc., are possible to estimate the subscriber positioning, but also new methods that take advantage of the properties of 5G networks, as it is the case of an emerging method known as Radio Frequency Pattern Match-ing (RFPM), especially for network operation at mmWaves [15]. RFPM, also called as fingerprinting, gathers a large database of Radio Frequency Patterns (RFPs) from signals imprinted over a set of receivers, and has fueled the interest of researchers [2,17], not just for positioning applications but also for UE identi-fication [3,10]. The collected signals could include the Received Signal Strength (RSS), the Channel Impulse Response (CIR), the Power Delay Profile (PDP), the Channel Transfer Function (CTF), the Frequency Coherence Function (FCF) [1], or the Angle-Delay Channel Amplitude Matrix (ADCAM), extracted from the Channel State Indicator (CSI) for massive Multiple Input Multiple Output (MIMO) - Orthogonal Frequency Division Multiplexing (OFDM) systems [21]. Furthermore, fusion techniques have been studied to improve positioning accu-racy [12]. Above studies have been performed with simulations, but a recent work studies the problem for indoor positioning via a test bench 5G-NR imple-mentation on FR1 band with the center frequency set to 2565 MHz and 10 MHz bandwidth, using the CSI extracted from the downlink synchronization signal block [18]. This work reports mean absolute errors between 2.14 and 2.81 m for one meter grid resolution. These values are similar to the reported accuracy between 2.20 m and 2.25 m for indoors RSS-based fingerprinting for one meter grid resolution and highly dense training scenarios [17]. However, RSS patterns can be degraded in FR1, reducing the accuracy of subscriber positioning. The practical approach based on CSI in [18] requires a highly complex signal pro-cessing. It is the paid price to achieve a high positioning resolution when errors around 14.9 m have been reported for outdoors in TOA-based positioning [18].

High resolution RFPM based on RSS has been achieved with the application of K-Nearest Neighbors (KNN), random forest, and Support Vector Machines (SVM); being the first method slightly better among these three [8]. Further, accuracy increases with the density of training fingerprints over the area of inter-est (AoI), so the direct application of RFPM over the whole AoI increases the computational burden with the inverse of the square of the grid resolution used to collect the RFPs. Therefore, [17] divides the AoI in smaller pieces to achieve an initial classification of the pattern into the most likelihood section using a coarse resolution grid and the four nearest neighbors for the application of KNN, and later proceeds with the RFPM using the highest available resolution within the selected section. This method achieved errors below 5 m in 85% of the cases and below 40 m in 95% of the cases, compared to errors below 5 m in 80% of the cases and below 40 m in 99% of the cases when KNN is directly applied to the whole AoI for the higher resolution grid. This 5% penalization in the location accuracy is the cost to reduce the computational burden to approximately one-third of the original. It has been proposed in [6] to automatically divide the AoI into clusters using the k-means clustering (k-means) algorithm, or any of its variants. Since the RFPs are affected by the channel dynamics due to environmental conditions and also by peculiar characteristics of RF circuitry and the height of the transmitters, dataset dimensions should grow to attend to these issues.

The k-means is widely favored due to its practicality, simplicity, and efficiency, and it has been successfully applied in many fields such as image segmentation, feature learning, and wireless sensor network management; but it exhibits several problems due to its random initialization and it is easily misled by outliers [14], so variations such as k-means++ or k-means# have been proposed to avoid erroneous initialization, and to detect and discard outliers respectively. k-means# preserves k-means simplicity and efficiency by maintaining its basic structure, altering only the centroid update step, and using a single global parameter value as a threshold to identify outliers. On the other hand, [11] proposes to incorporate a genetic algorithm to k-means to improve its capabilities to reject outliers, and provides results showing that genetic k-means outperforms the original method. Several algorithms propose modifications to k-means to accelerate the clustering process [13,20]. For example, [13] provides a method to avoid certain distance recalculations in k-means to alleviate the computational burden, whilst [20] includes modifications into k-means taken from the k-medoid algorithm to facilitate the procedure. Research in [7] proposes some modifications to k-means, so this new algorithm can obtain clusters in preferred sizes. This proposal has a relatively lesser chance of getting trapped in local minima as the centroids are selectively initialized with data points from each cluster using prior knowledge. Therefore, this algorithm can produce more reasonable results compared to the standard k-means.

The contributions of this work are the following: 1) It provides a realistic simulation environment suitable for studying the subscriber location issue in 5G mmWave networks. RSS fingerprints are used for simplicity, but more complex signals could be extracted from this Matlab® model to test other positioning architectures [21]. 2) It studies the application of ML clustering algorithms for the automatic identification of environment features using a light computational burden procedure and their high-resolution interpolation to adapt these results to KNN techniques to achieve subscriber position estimation using RFPM. 3) It studies the effect of operation frequency on RFPM-based positioning.

Figure 1 summarizes in a flowchart the methodology employed throughout this research.

This document is organized as follows: Section 1 makes a brief review of the RFPM issue as a means to perform wireless subscriber radio-location, and from the potential of k-means and other clustering algorithms to discover the required environment features prior to performing a high-resolution matching, with the help of ML KNN method; Section 2 introduces the simulation settings for the

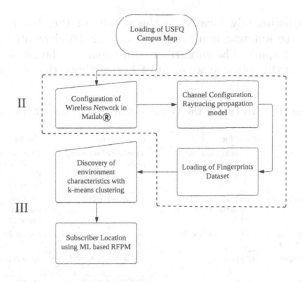

Fig. 1. Methodology Flowchart.

deployed wireless network, the fingerprints dataset employed for the application and referential supervised ML methods; Section 3 presents the k-means clustering algorithm and the proposed smarter method used to achieve radio positioning, Sect. 4 shows the new positioning results and compares them with previous research; and Sect. 5 summarizes the main observations and conclusions of this work and also suggests some topics for future research.

2   Deployed Wireless Network, RFPM, and Supervised ML Methods

A distributed massive MIMO (DM-MIMO) wireless system [2] has been virtually deployed along the campus of Universidad San Francisco de Quito (USFQ) in Ecuador using Open Street Maps (OSM) and Matlab®. This kind of system uses large-scale antenna arrays at the base station (BS). Signals transmitted from User Equipment (UE) are highly affected by multipath due to surrounding scattering, so the received fingerprints at the BS are characteristic of the local environment around these subscribers.

2.1   Network Deployment Setup

A dense urban scenery in the 5G enhanced Mobile Broadband (eMBB) has been set up by taking into account current specifications for FR2 bands as summarized in Table 1 and using the combination method to characterize the 3D antenna patterns. Deployed transmitters and receivers have been set up as in [19]. The testing area is a rectangular location of 500×300 m^2, as shown in Fig. 2. The eNodeB is approximately located in the middle of the AoI to maximize the coverage, and the antenna is a square array with 16 elements where neighbor elements are 5 m apart. The subscriber equipment will be randomly located at any point within this place at a height of 1.50 m above the terrain surface.

Table 1. Settings for the Dense Urban-eMBB test environment

Parameters	Config.	Parameters	Config.
Evaluation Carrier frequency	1 layer (macro) at 30 GHz	*Total transmission power for TRxP*	40 dBm with a bandwidth 80 MHz EIRP restricted to be lower than 73 dBm
Simulation bandwidth	80 MHz for TDD, 40 MHz + 40 MHz for FDD	*UE power class*	23 dBm, EIRP restricted to be lower than 43 dBm
Number of antenna elements per TRxP	Up to 256 Tx/Rx	*Number of antenna elements of UE*	Up tp 32 Tx/Rx
BS antenna height (gNB)	25 m	*UE antenna height*	At outdoors: 1.5 m
BS Noise Figure	7 dB	*UE Noise Figure*	10 dB
Gain of BS antenna element	8 dBi	*Gain of UE antenna element*	5 dBi

2.2   Propagation Model, RFPs and ML-KNN

The wireless channel has been set up as in [19], by using the Ray Tracing (RT) propagation model, limited to 4 reflections due to the received signal power that is often well characterized by the first two arriving paths. RSS measures gathered on this array become the RFPs (fingerprints) used for subscriber location. A new RFP is collected at the BS array for each new transmitter position on a grid deployed over the AoI.

For simulations in this work, two datasets were used. The former corresponds to a collection of 6000 RFPs measured from a 5 m resolution grid, and the latter to 660 RFPs from a 15 m resolution grid. In addition, methods in [17] will be compared to our proposal. The referential methods for radio-location were a) One-step KNN (1S-KNN), b) Two-step KNN (2S-KNN), and c) Supervised clustering followed by 2S-KNN (SC+2S-KNN). The main piece of all these procedures is the KNN algorithm. It refers to a supervised method that operates over a group of points and classifies any new point in relation to a distance metric toward samples belonging to predefined classes in the near neighborhood. This method requires training and a prior interpretation of the reference data [17]. The most computationally efficient method of this group is SC+2S-KNN. Our proposal will modify it to achieve a smarter one.

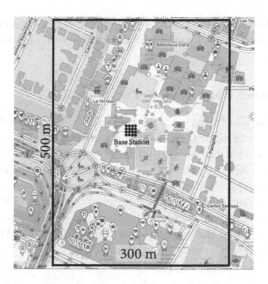

Fig. 2. Area of Interest (AoI).

3   Unsupervised Clustering

SC+2S-KNN uses a completely supervised approach. It arbitrarily divides the
AoI into three regions and assigns each new RFP to one of them using KNN.
The aim of our proposal is to include intelligence in the clustering process with
the use of an unsupervised clustering (USC) ML algorithm such as k-means++.
Dividing the AoI into smaller regions has a positive impact on the computational
burden as mentioned in the introduction, and a USC ML algorithm has the
ability to easily recognize data similarity, reducing the clustering arbitrariness
and automatically identifying the outliers.

3.1   The K-means Algorithm

k-means provides an automatic solution for the classification problem and does
not require the setting of arbitrary borders for the collected information to pro-
ceed. In other words, k-means provides a technique to categorize data. The k-
means algorithm partitions a given set of data points into k clusters and assigns
each data point to one of them. The initial centroids are arbitrarily selected from
the dataset and are adjusted iteratively by the algorithm until there is no longer
a significant change between new and previous results and an optimized vector
of centroids is finally found. In this work, Matlab® k-means++ implementation was used since it optimizes the initialization of the selection of the cluster
centroids, achieving better results than the original k-means algorithm. [4].

Additionally, Matlab allows for further optimization of the k-means++ algorithm by performing multiple replicas or realizations. When executing the k-
means command, it is possible to specify the number of times n, the process will

be initialized and conducted. Therefore, the data will be clustered n times, with a new initial centroid and a local minimum for the sum of distances determined in each run.

3.2   Unsupervised Clustering + Two-Step KNN

Our Unsupervised Clustering + Two-Step KNN (USC+2S-KNN) method is based on the k-means algorithm. This method performs k-means one time to estimate the region that the subscriber RFP belongs to and concludes with the application of 2S-KNN into the previously obtained region to eventually estimate the subscriber position within the high 5m resolution grid. The method was subsequently a) optimized for the k number of regions in k-means clustering and b) optimized for the K number of neighbors in KNN with the best radio positioning results. Using the optimized parameters, the new ML method is then c) compared with the referential ML methods.

Training Stage. Figure 3 shows the methodology for the training stage, in which both datasets of 660 RFPs and 6000 RFPs were employed. First, the 660 RFPs dataset was subjected to the k-means clustering algorithm to divide the AoI into $k + 1$ regions, with $k = 3$. The additional non-identified region (NIR) was included by the algorithm when it identified spurious cases with the weakest RSS levels. Then, the resulting k-means clustering 15 m resolution matrix was expanded to a higher 5 m resolution mask using the Kronecker product between the original data and a 3×3 all ones matrix [5]. This high-resolution mask, depicted in Fig. 4, labels each fingerprint into its corresponding region within the 6000 RFPs dataset. In the final step, this dataset jointly with the corresponding UE's reference positions on the grid were used to train the KNN algorithm along each labeled region. For this step, NIR was excluded due to its null contribution.

Implementation/Testing. For the implementation stage, a new set of 300 RFPs provided by random UE positions on the AoI was employed. Each test RFP is first associated with the nearest k-means cluster centroid to find the region to which it belongs. Then, the KNN algorithm was implemented to match this RFP to the one with the nearest reference transmitter position. The KNN algorithm is applied strictly within the previously selected labeled region.

4   Results

The results for the set of 300 RFPs provided by random UE positions on the AoI using the new ML method (USC + 2S-KNN) are shown as follows.

4.1   Optimal Number of Clusters for K-Means

To determine the optimal number of clusters, the USC+2S-KNN method was evaluated for different values of k clusters ranging from 3 to 10, and keeping

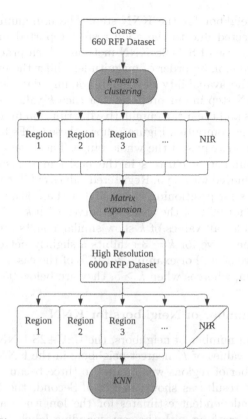

Fig. 3. Training Stage Methodology Flowchart.

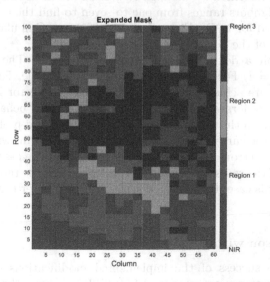

Fig. 4. High-Resolution Expanded Clustering Mask.

just one nearest neighbor for the KNN step. The minimum value of k within this range was selected due to the good results reported in [17] with a three region split and the use of SC+2S-KNN method, which practically reduces the computational burden in an order of magnitude; whilst the maximum value was chosen to ensure the availability of enough training data within the resulting regions for the KNN step in our proposal. For each k value, the experiment was performed 50 times and the mean cumulative distribution function (CDF) for the positioning error was computed. Figure 5 shows these CDFs for k values of three, five, and ten, representatives of the whole range. These values show the impact of changing the number of clusters in the positioning results. No significant differences were achieved for $k \geq 5$. Registered values exhibit fluctuations around 1% for these curves for positioning errors higher than 60 m. The position error is lower than 60 m for 96% of the cases for curves with $k \geq 5$.

Further, curves for all values of k show similar results for the position error below 30 m, but the curve for $k = 3$ exhibits a slightly better behavior for any position error above 30 m. For example, for 97% of the cases, when $k = 3$, errors are lower than 45 m, whereas when $k \geq 5$, they are below 95 m.

4.2   Optimal Number of Neighbors for KNN

To find the optimal number of neighbors, the USC+2S-KNN method was evaluated for different values of K nearest neighbors in the KNN algorithm.

First, the number of regions was limited to three because this configuration exhibited the best results as shown in Fig. 5. Second, the KNN is performed twice, providing independent estimates for the longitude and latitude of the subscriber position working with the corresponding labeled portion of the 6000 RFPs dataset as informed by the high-resolution mask. In this second part, the number of K neighbors ranges from one to seven to find the optimal number of neighbors required for KNN. Figure 6 shows the mean Cumulative Distribution Function (CDF) of the positioning error for these values.

Curves exhibit a degraded behavior as K increases. The best results are achieved for $K = 1$. For $K \geq 5$ differences are negligible. For $K = 1$, errors lower than 30 m are achieved in 95% of the cases. This error increases to 43 m for $K = 2$ and deteriorates up to 85 m when $K = 4$. This behavior implies that this method relies on the availability of a large dataset and prefers a direct comparison between the current RFP and each stored fingerprint within a particular region. For $K \geq 5$, errors are lower than 7 m for 69% of the cases and lower than 100 m for 92% of the cases. Positioning error depends on the resolution of the reference grid. In this case, the expected errors will be greater than approximately 3 m.

4.3   Comparison with Previous Works

To quantify the success of the implemented modifications on the proposed method, it is necessary to compare USC+2S-KNN, with the former methods described in section II [17].

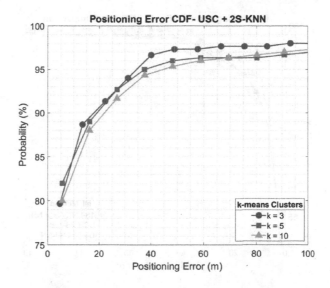

Fig. 5. CDF for k-means + 2S-KNN RFPM Technique.

Figure 7 compares the CDFs of the positioning error for the optimized USC+2S-KNN with $k = 3$ clusters and $K = 1$ nearest neighbor, with 1S-KNN, 2S-KNN and SC+2S-KNN methods.

Our new USC+2S-KNN method exhibits similar behavior with positioning errors lower than 16 m for 91% of the cases if compared to SC+2S-KNN but

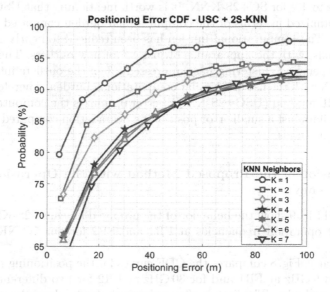

Fig. 6. CDF for USC + 2S-KNN in terms of Number of Neighbors.

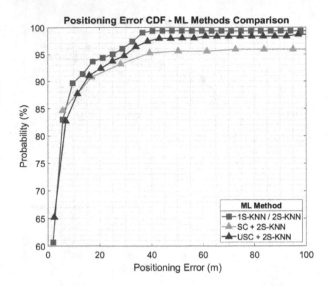

Fig. 7. CDFs of different Machine Learning Methods.

outperforms it by reaching an error lower than 42 m for 98% of the cases, whilst SC+2S-KNN just achieves this precision for 96% of the cases. However, 2S-KNN achieves errors lower than 40 m for 99% of the cases. Further, these results exhibit significant positioning error differences in the worst cases. Just 1% of the cases exhibit errors greater than 40 m when 1S-KNN or 2S-KNN are used. This portion increases to 2% for the new proposed USC+2S-KNN method and corresponds to 4% for SC+2S-KNN. It is worth mentioning that USC+2S-KNN is highly optimized in terms of the computational burden compared to 1S-KNN or 2S-KNN. The computational burden has been reduced to nearly one-ninth of these methods, with the appearance of just 1% of new outliers. The incorporation of unsupervised clustering has suppressed 2% of the outliers introduced by SC+2S-KNN with the use of a similar computational burden. Therefore, the new proposed ML method (USC+2S-KNN) better optimizes the computational burden and exhibits just a small error positioning degradation compared to 1S-KNN or 2S-KNN methods.

4.4   Behavior of the Proposed Method with the Operation Frequency

Figure 8 and Fig. 9 show the behavior of the proposed kmeans+2S-KNN method for different operation frequencies at FR1 and FR2 for our 5G-NR referential system.

Particularly, Fig. 8 compares the CDF curves for the positioning precision for 3.7 GHz and 6 GHz at FR1 and for 30 GHz at FR2 for two different array configurations at the BS. The first configuration includes 16 elements at the square antenna array, whilst the latter includes just 9 elements. The first observation

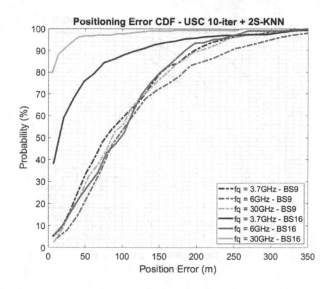

Fig. 8. CDFs for k-means+2S-KNN at FR1 and FR2 and two different sensor array configurations.

refs to the strong precision degradation associated to the array dimension. For example, the positioning error is less than 5 m for 80% of the cases, below to 50 m for 95% of the cases and lower than 120 m for 98% of the cases at FR2 for the 16-elements array. This precision degrades to less than 70 m for 70% of the cases, below to 175 m for 90% and lower than 265 m for 99% of the cases for the shorter 9-elements array. Another important observation is that positioning error deteriorates for FR1, especially for 6 GHz. For example, at 3.7 GHz, error is less than 60 m for the 80% of the cases, below to 120 m for 90% of the cases and lower than 265 m for 99% of the cases; whilst, at 6 GHz, it degrades to less than 60m for 30% of the cases, below to 130 m for 70% of the cases, lower than 180 m for 90% of the cases and lower than 265 m for the 99% of the cases. Finally, it is important to comment that differences are very low when the shorter array is used instead of the original. For example, error is lower than around 130 m for the 70% of the cases for the three considered frequencies. Furthermore, CDF curves for 30 GHz and 3.7 GHz with 9-elements array are quite close to the curve for 6 GHz for the 16-elements array for higher positioning errors. The worst CDF corresponds to 6 GHz with errors lower than 250 m for 90% of the cases, and errors below to 350 m for the 99% of the cases.

Figure 9 shows the CDFs of the positioning error for different number of replicas of the KMC algorithm. It is interesting to note as the error reduces especially for curves at 3.7 GHz when the number of replicas changes from 1 to 5. No improvement is achieved for a number of replicas higher than 5. Results exhibited in Fig. 8 and commented above correspond to the best results achieved when 5 replicas are used.

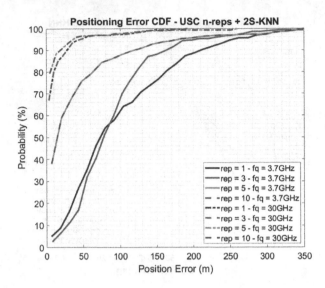

Fig. 9. CDFs for k-means+2S-KNN for several algorithm replicas at FR1 and FR2.

Results not exhibited in this paper for a 9-element array receiver at FR2 shows a significantly larger degradation than the 16-element array curves. Furthermore, when varying the number of clusters, it was determined that the best results are obtained when the number of clusters are $k = 3$ and $k = 5$ for the 16-element array, which correspond to the best results exhibited in Fig. 8. Likewise, the best results for the 9-element array were obtained when $k = 3$, $k = 5$, and $k = 7$, which also correspond to those from Fig. 8. For the 9-element array, the number of clusters seems to have a smaller impact in the results, thus being $k = 10$ the only curve that differs from others showing a bigger positioning error.

5 Conclusions and Future Work

This paper exhibits the deployment of a realistic simulation for a 5G-NR DM-MIMO wireless network operating at FR1 and FR2, and tests a USC+2S-KNN RFPM method developed to obtain the subscriber positioning of users located within the AoI. This new method incorporates the k-means++ algorithm to the former 2S-KNN RFPM method by implementing an unsupervised ML classifying algorithm that automatically divides the AoI into smaller clusters grouped by similarity features with higher success in the outlier suppression compared with SC+2S-KNN RFPM method used as reference. At FR2, this new approach introduced just a 1% of new outliers compared with the original 2S-KNN RFPM method with approximately one-ninth of the computational burden, reducing the number of outliers to one-third of the achieved by the SC+2S-KNN RFPM with a similar computational burden. The new method accuracy degrades about 2 m in terms of the mean positioning error as compared to 2S-KNN RFPM used as a reference, whose mean error is 7.5 m as shown in [2]. However, results performed with a shorter 9-element array collapse, and similarly good results obtained at FR2 suffer an important degradation at FR1, especially for an operation frequency of 6GHz. Although achieved results for 3.7 GHz show positioning errors lower than 50 m for 75% of the cases, further research is still required to determine appropriate methods to enhance positioning precision for these frequency bands and when the array dimension reduces. As future work, due to the successful performance of the unsupervised clustering for this application, which can significantly reduce the engineering task and can easily be expanded toward different environments, new clustering algorithms, such as k-means# [11,14] may be explored to further reduce the outliers and positioning errors. It would be interesting to assess the impact of using higher resolution on low SS coverage regions and their boundaries as a means of reducing the probability of misclassification and improving positioning accuracy. Moreover, the self-identification of the optimum number of clusters is possible [16], and could possibly benefit the physical deployment of this research. It is also important to evaluate the robustness of these methods by introducing certain randomness to our scenery such as UE's height, and meteorological conditions. Furthermore, research relative to the incorporation of new 5G specialized signals is also possible [18] to improve positioning accuracy. In addition, some empirical data retrieved from specialized network equipment could be analyzed and compared to our results.

References

1. Al Khanbashi, N., et al.: Real time evaluation of RF fingerprints in wireless LAN localization systems. In: 2013 10th Workshop on Positioning, Navigation and Communication (WPNC), pp. 1–6 (2013). https://doi.org/10.1109/WPNC.2013.6533273
2. Al-Rashdan, W.Y., Tahat, A.: A comparative performance evaluation of machine learning algorithms for fingerprinting based localization in DM-MIMO wireless systems relying on big data techniques. IEEE Access **8**, 109522–109534 (2020). https://doi.org/10.1109/ACCESS.2020.3001912. https://ieeexplore.ieee.org/document/9115664/
3. Al-Shawabka, A., et al.: Exposing the fingerprint: dissecting the impact of the wireless channel on radio fingerprinting. In: IEEE INFOCOM 2020 - IEEE Conference on Computer Communications, pp. 646–655 (2020). https://doi.org/10.1109/INFOCOM41043.2020.9155259
4. Arthur, D., Vassilvitskii, S.: K-Means++: The Advantages of Careful Seeding, pp. 1027–1035. Society for Industrial and Applied Mathematics, USA (2007)
5. Atteya, M.: Kronecker product with applications. J. Algebra Appl. **1**, 2016–2017 (2016). https://doi.org/10.14419/jaa.v1i1.35
6. Bai, S., Wu, T.: Analysis of k-means algorithm on fingerprint based indoor localization system. In: 2013 5th IEEE International Symposium on Microwave, Antenna, Propagation and EMC Technologies for Wireless Communications, pp. 44–48 (2013). https://doi.org/10.1109/MAPE.2013.6689952

7. Ganganath, N., Cheng, C.T., Tse, C.K.: Data clustering with cluster size constraints using a modified k-means algorithm. In: 2014 International Conference on Cyber-Enabled Distributed Computing and Knowledge Discovery, pp. 158–161 (2014). https://doi.org/10.1109/CyberC.2014.36

8. Han, X., He, Z.: A wireless fingerprint location method based on target tracking. In: 2018 12th International Symposium on Antennas, Propagation and EM Theory (ISAPE), pp. 1–4 (2018). https://doi.org/10.1109/ISAPE.2018.8634177

9. ITU: Report ITU-R M.2412-0: Guidelines for evaluation of radio interface technologies for IMT-2020 (2017). https://www.itu.int/dmspub/itu-r/opb/rep/R-REP-M.2412-2017-PDF-E.pdf

10. Kaminski, J.S., Stein, D., Weitzen, J.: Removing the RF fingerprint: a least squares approach to compensate for a device's hardware impairments. In: MILCOM 2022-2022 IEEE Military Communications Conference (MILCOM), pp. 374–378 (2022). https://doi.org/10.1109/MILCOM55135.2022.10017915

11. Kapil, S., Chawla, M., Ansari, M.D.: On k-means data clustering algorithm with genetic algorithm. In: 2016 Fourth International Conference on Parallel, Distributed and Grid Computing (PDGC), pp. 202–206 (2016). https://doi.org/10.1109/PDGC.2016.7913145

12. Li, X., Chen, L., Zhou, X., Ruan, Y., Liu, Z.: Fusion of 5G carrier phase ranging and 5G CSI fingerprinting for indoor positioning. In: 2022 14th International Conference on Wireless Communications and Signal Processing (WCSP), pp. 1–5 (2022). https://doi.org/10.1109/WCSP55476.2022.10039441

13. Na, S., Xumin, L., Yong, G.: Research on k-means clustering algorithm: an improved k-means clustering algorithm. In: 2010 Third International Symposium on Intelligent Information Technology and Security Informatics, pp. 63–67 (2010). https://doi.org/10.1109/IITSI.2010.74

14. Olukanmi, P.O., Twala, B.: K-means-sharp: modified centroid update for outlier-robust k-means clustering. In: 2017 Pattern Recognition Association of South Africa and Robotics and Mechatronics (PRASA-RobMech), pp. 14–19 (2017). https://doi.org/10.1109/RoboMech.2017.8261116

15. del Peral-Rosado, J.A., Raulefs, R., López-Salcedo, J.A., Seco-Granados, G.: Survey of cellular mobile radio localization methods: from 1G to 5G. IEEE Commun. Surv. Tutor. **20**(2), 1124–1148 (2018). https://doi.org/10.1109/COMST.2017.2785181

16. Rajeswari, K., Acharya, O., Sharma, M., Kopnar, M., Karandikar, K.: Improvement in k-means clustering algorithm using data clustering. In: 2015 International Conference on Computing Communication Control and Automation, pp. 367–369 (2015). https://doi.org/10.1109/ICCUBEA.2015.205

17. René, J.E., Salazar, A., Beltrán, K., Caisaluisa, O.: Subscriber location in 5G mmWave networks - machine learning RF pattern matching. In: 2022 IEEE International Autumn Meeting on Power, Electronics and Computing (ROPEC), vol. 6, pp. 1–6 (2022). https://doi.org/10.1109/ROPEC55836.2022.10018626

18. Ruan, Y., et al.: iPos-5G: indoor positioning via commercial 5G NR CSI. IEEE Internet Things J. **10**(10), 8718–8733 (2023). https://doi.org/10.1109/JIOT.2022.3232221

19. Salazar, A., Arévalo, G.V., Játiva, R.: Propagation, blockage and coverage evaluation in 5G urban wireless networks. In: 2021 Global Congress on Electrical Engineering (GC-ElecEng), Valencia, Spain, pp. 55–60. IEEE (2021). https://doi.org/10.1109/GC-ElecEng52322.2021.9788150. https://ieeexplore.ieee.org/document/9788150/

20. Shah, S., Singh, M.: Comparison of a time efficient modified k-mean algorithm with k-mean and k-medoid algorithm. In: 2012 International Conference on Communication Systems and Network Technologies, pp. 435–437 (2012). https://doi.org/10.1109/CSNT.2012.100

21. Sun, X., Wu, C., Gao, X., Li, G.Y.: Fingerprint-based localization for massive MIMO-OFDM system with deep convolutional neural networks. IEEE Trans. Veh. Technol. **68**(11), 10846–10857 (2019). https://doi.org/10.1109/TVT.2019.2939209

Robotic Arm Based on Artificial Intelligence for Learning Braille as an Inclusive Tool in Educational Environments

Santiago S. Puentes G., Maria C. Moreno, Brayan Daniel Sarmiento,
and Oscar J. Suarez$^{(\boxtimes)}$

Mechatronics Engineering Department, University of Pamplona, Pamplona, Colombia
oscar.suarez@unipamplona.edu.co

Abstract. In the last few years, applications within the field of service robotics have increased exponentially due to their potential to assist people in a wide variety of tasks related to agriculture, education, mining, healthcare, security, and defense sectors. Therefore, this chapter introduces the design and development of a robotic arm with four degrees of freedom (DoF) intended for implementation in educational environments as an interactive robotics tool for teaching the Braille system. This project arose from the need for inclusive and accessible education, using robotics to bring Braille to a broader audience and encourage collaborative learning. Integrating kinematics, dynamics, as well as cubic spline interpolators, ensures accurate and efficient movements. Additionally, an artificial neural network model was developed and trained to calculate the kinematic inverse of the robotic arm, demonstrating high accuracy and reliability. The system allows users to trace letters or words using a graphical user interface on mobile and computing devices. Additionally, a Convolutional Neural Network (CNN) was implemented to facilitate data acquisition and perform translation of Braille characters into Spanish.

Keywords: Service Robotics · Educational Environments · Convolutional Neural Network · Braille

1 Introduction

Robotics has become a fundamental tool in different fields, including education and assistance for the visually impaired [1]. STEM education (Science, Technology, Engineering, and Mathematics) has been widely used to foster interdisciplinary learning and skill development in areas such as technology, engineering, and art, among others [2]. SMART, STEM, and STEAM learning technologies

This work is supported by University of Pamplona, Pamplona, Colombia.

enable the training of highly skilled specialists for the digital society by integrating robotic systems into education [3]. Applying active methodologies based on mechatronics and robotics can integrate STEAM competencies into current curricula [4]. A systematic review study shows that using physical and robotic devices with collaborative methodologies has succeeded in STEAM education, demonstrating the potential of robotics and Mechatronics to engage students and enhance learning [5].

Integrating robotics into education has shown promise in improving creativity, teamwork, and knowledge acquisition in specific areas [6–8]. In addition, educational robotics has benefited child cognitive development and the inclusion of students with disabilities [9,10]. Educational robotics has proven to be a valuable tool in inclusive education for children and adolescents with disabilities. It was found that most participants in these environments improved their skills, facilitating their inclusion in educational settings in some cases. These findings support the need to implement educational robotics practices to achieve successful inclusion [11].

In the field of assistance for people with visual impairments, different projects have explored the use of robotics and technology to improve their quality of life and mobility. Initiatives such as the smart cane with sensors and Global Positioning System (GPS) [12], classroom robots for children with mixed visual abilities [13], Braille character detection for visually impaired people using neural networks in an educational robot [14], and LEGO Braille Bricks have also emerged to teach the Braille system to children with visual limitations and normal visual acuity through play [15]. These examples guarantee adequate mobility and inclusion in the classroom but need to be educational tools that facilitate learning for people with visual impairment.

Based on the considerations mentioned above, this chapter presents the design, prototyping, and experimental testing of a robotic arm with four degrees of freedom intended for learning Braille for individuals with normal visual acuity by tracing letters or words using an easily accessible graphical user interface on mobile and computing devices. Furthermore, a convolutional neuronal network has been implemented to facilitate data acquisition and trajectory creation. Additionally, translate the Braille characters into Spanish through identification for improving education and the inclusion of people with visual impairment, thus promoting equal learning opportunities [16,17].

The remainder of this chapter is organized as follows. Section 2 summarizes the concepts implemented during the study. Section 3 provides the kinematic and dynamic model, the cubic spline interpolation, the mechanical and electronic design, the development of a personalized audio system and an artificial neural network for the execution of learning routines and detection of Braille characters, as well as the integration of a graphical user interface for computer and mobile devices. Section 4 describes the experimental results of the proposed prototype. Finally, Sect. 5 outlines the authors' conclusions and future work.

2   Preliminaries

This section presents a brief review on Convolutional Neural Networks, YOLO technology, Educational and Inclusive Robotics, Graphical User Interface, and Personalized Audio Resources on Inclusive Learning.

2.1   Convolutional Neural Networks (CNNs)

In the realm of aiding visually impaired individuals, two distinct studies shed light on the pivotal role of Convolutional Neural Networks (CNNs) in enhancing accessibility. The first study focuses on Braille block recognition, recognizing the challenges faced by visually impaired individuals in locating these essential tools. Utilizing a camera and CNNs, this research successfully achieved a commendable test accuracy of approximately 94%, providing a robust solution to guide visually impaired individuals to Braille blocks [18]. Meanwhile, the second study delves into Arabic Braille numeral recognition, addressing the complexity of memorizing dot arrangements within characters. Employing a Residual Network (ResNet) variation of CNNs, this research impressively attained a recognition accuracy of 98% for Arabic Braille numerals under varying lighting and distance conditions, further emphasizing the potential of CNNs in transforming Braille content into plain text and speech, significantly enhancing accessibility for the visually impaired [19].

2.2   You Only Look Once (YOLO)

Implementing the "You Only Look Once" (YOLO) technology represents a significant advancement in searching for inclusive solutions for visually impaired people. Like educational robotics, YOLO plays a crucial role by using image recognition and Deep Learning techniques to provide 3D auditory information about the location of objects and obstacles in front of the user [20]. This revolutionary approach seeks to detect objects, recognize their nature, and convey that information effectively to the user. By incorporating the perspective of a severely visually impaired person through an interview, a more comprehensive and focused approach to the user's needs has been achieved, underlining the importance of inclusion and accessibility in applying cutting-edge technology to improve the quality of life for the visually impaired.

2.3   Educational and Inclusive Robotics

As exemplified in this project, educational robotics played a pivotal role in augmenting cross-disciplinary learning. The project harnessed robotic devices to cultivate fundamental competencies such as logical reasoning, problem-solving, programming, and engineering skills. The significance of these contributions to skill development cannot be overstated. Furthermore, the project underscored the potential of educational robotics to champion inclusive and collaborative learning environments. In the realm of special education, its primary focus centered on delivering innovative opportunities tailored to students with diverse learning needs. Integrating robotics into pedagogical approaches effectively yielded engaging and accessible learning experiences, transcending barriers posed by disabilities or unique educational challenges, as documented in references [6, 7]. Customizing educational robotics activities aligning them with varying learning styles and abilities, ultimately fostered an inclusive educational milieu. Every student actively participated in this environment, capitalizing on the learning process to the fullest extent.

2.4   Graphical User Interface (GUI)

In educational technology, the significance of a user-friendly interface cannot be overstated, [21, 22]. Just as educational robotics can enhance cross-disciplinary learning and inclusivity in the learning environment, a well-designed user interface in software or digital tools can serve as a bridge to intuitive and accessible learning experiences. Just as educational robotics tailored activities to diverse learning needs, a thoughtfully crafted interface can accommodate various user profiles, ensuring that information is presented clearly and comprehensibly. The parallels between these studies and interface design principles emphasize that user-friendliness is essential in facilitating effective learning. By harnessing the power of intuitive design, educators can break down barriers, making educational content more engaging and accessible to all, thus maximizing the learning potential of each individual.

2.5   Personalized Audio Resources on Inclusive Learning

Customized audio materials have emerged as a powerful tool for enhancing the learning experience, as evidenced by the research discussed in the articles [23–26]. In [23], the use of personalized audio feedback was shown to positively impact students' engagement and motivation in educational robotics, transcending barriers and fostering inclusivity. Similarly, in [24], the incorporation of personalized audio instructions in the context of language learning showcased significant improvements in pronunciation and comprehension. [25] delved into the importance of addressing individual learning needs, and personalized audio interventions were identified as a valuable strategy for improving neurosensorial learning. Furthermore, in [26], personalized audio messages enhanced the understanding of complex mathematical concepts. Collectively, these studies underscore the pivotal role of customized audio materials in catering to diverse learning needs, breaking down barriers, and creating inclusive learning environments. Just as educational robotics and user-friendly interfaces have the potential to make learning more accessible, personalized audio resources offer a dynamic avenue to optimize the learning process for each individual, ultimately maximizing their educational potential.

3 Materials and Methods

This section introduces the general design of the robotic structure and describes the logical architecture of the system and the integration of a neuronal network for long-term robot autonomy.

3.1 Forward and Inverse Kinematics

Solving the forward kinematics problem is reduced to finding a Homogeneous Transformation Matrix (HTM) related to the position and orientation of the robot end-effector, which takes the robot base as a fixed reference system. The Denavit-Hartenberg (D-H) algorithm is implemented to determine the forward kinematic model. The algorithm's results are used to obtain the analytical results of the robot's coordinate system as can be seen in Fig. 1, which will be used to complete the D-H matrix.

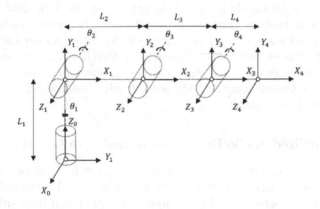

Fig. 1. Robot coordinate system.

Table 1 presents the robot's geometric parameters according to the Denavit-Hartenberg convention, where i represents the joint number, θ_i represents the angle with respect to the x_i and x_{i+1} axis, d_i represents the distance between the z_i axes, a_i represents the distance along the x_i axis, and α_i refers to the angle between the z_i and z_{i+1} axes.

For the inverse kinematics problem, the kinematic decoupling method was used, whereby a desired final position and orientation is given to establish the coordinates of the counting point in the final three axes, obtaining the first three articular variables. The value of the fourth articular variable is obtained based on the orientation data and the previously calculated articular variables. To verify the correct implementation of the inverse kinematics of the four-degree-of-freedom robot, simulations and tests were performed in MATLAB environment.

Table 1. Manipulator kinematics Parameters

D-H Parameters	θ [°]	d [mm]	a [mm]	α [°]
Link 1	q_1+90	L_1	0	90
Link 2	q_2	0	L_2	0
Link 3	q_3	0	L_3	0
Link 4	q_4	0	L_4	0

3.2   Dynamic Analysis and Forces

Accurate path tracking is essential in robotic manipulation. Although industrial robots address this problem through rigidity and high-performance hardware, affordable and compatible robots require advanced control to achieve accurate position tracking [27]. The Lagrange-Euler formulation based on energy considerations using the robot's inverse dynamic model is proposed to select the optimal servomotors. The mathematical expression for the Lagrangian formulation is given by Eqs. (1) and (2).

$$\frac{d}{dt}\frac{\partial \mathcal{L}}{\partial \dot{q}_i} - \frac{\partial \mathcal{L}}{\partial q_i} = \tau, \tag{1}$$

$$\mathcal{L} = K - U, \tag{2}$$

where:

$\mathcal{L}$: Lagrangian Function.
q_i : Articulate Coordinates.
τ : Applied Pair Vectors.
K : Kinetic Energy of the System.
U : Potential Energy of the System.

Implementing the Lagrange-Euler algorithm contemplates parameters of the mechanical structure related to the centers of inertia, centers of mass, and masses. For this, it was necessary to analyze the materials and characteristics of each robot link using computer-assisted programs. Then, the Lagrange-Euler recursive algorithm was applied. This can be described as follows in Eq. (3).

$$\tau = D(q)\ddot{q} + H(q,\dot{q}) + C(q) + F_v\dot{q}, \tag{3}$$

where:

D : Inertial Matrix.
H : Vector of Coriolis and Centrifugal Forces.
C : Gravitational Force Vector.
F_v : Robot's Friction Coefficient Matrix.

Subsequently, a simulation was developed in the MATLAB-Simulink environment to evaluate the robot in the most demanding configuration, applying parameters based on trapezoidal velocity profiles (TVPs). Figure 2 presents the Simulink scheme for calculating the maximum torques required to facilitate the selection of the servomotors in the system.

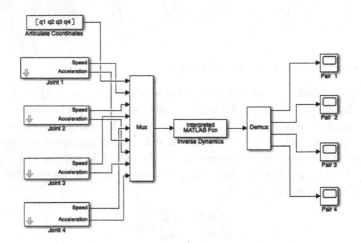

Fig. 2. Calculation for the maximum pairs required by the robot.

3.3   Cubic Spline Interpolation

In order to present continuity in the paths that each articulation must generate, it was necessary to use a polynomial of degree 3 that joins each pair of adjacent points. This enabled imposing four boundary conditions, two position conditions, and two velocity conditions. Note that the speed values must be calculated in advance for each point. The mathematical expression of the path that joins two adjacent points (q^{i-1}, q^i) is described below in Eq. (4).

$$q(t) = a + b(t - t^{i-1}) + c(t - t^{i-1})^2 + d(t - t^{i-1})^3, \qquad (4)$$

where q represents the articulate coordinates and t represents the time instants of passage. Consequently, each parameter that makes up the previous equation is defined through Eqs. (5) (6) (7) (8), and (9) for $t^{i-1} < t < t^i$ condition.

$$a = q^{i-1}, \qquad (5)$$

$$b = \dot{q}^{i-1}, \qquad (6)$$

$$c = \frac{3}{T^2}(q^i - q^{i-1}) - \frac{2}{T^2}\dot{q}^{i-1} - \frac{1}{T^2}\dot{q}^i, \tag{7}$$

$$d = -\frac{2}{T^3}(q^i - q^{i-1}) + \frac{1}{T^2}(\dot{q}^{i-1} + \dot{q}^i), \tag{8}$$

$$T = t^i - t^{i-1}. \tag{9}$$

This method can be used to evaluate the robot's accuracy and efficiency on specific tasks, and the model can be adjusted if necessary to optimize performance. Figure 3 presents a simulation using a cubic splines interpolator in MATLAB to generate the word "Hello" in Braille with a number of points equal to 10 during the first and second sampling.

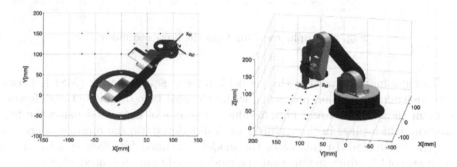

Fig. 3. Simulation of the robot in MATLAB.

3.4   Artificial Intelligence for Long-Term Autonomy

A deep neural network model using the Keras API of TensorFlow was developed for long-term robot autonomy. The model's architecture consists of six fully connected layers with ReLU activation functions and one final layer with a linear activation function, which predicts a constant value. The input layer has three neurons corresponding to the dataset's three features.

Furthermore, the number of neurons in the hidden layers progressively increases from 128 to 512 and then decreases back to 128. This architecture was chosen to allow the model to learn complex and nonlinear patterns in the data by capturing both low and high-level features. The optimizer used in this model is Adam, a popular and effective deep-learning optimization algorithm. The optimizer's learning rate is set to 0.0001 a lower value than the default learning rate, and was chosen to help the optimizer better converge and avoid overshooting the global minimum of the loss function. The comparison between precision and loss for the training and validation sets is presented in Fig. 4.

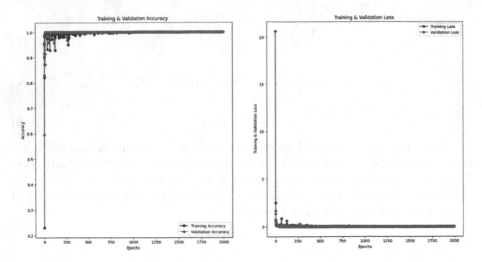

Fig. 4. Graphs resulting from the proposed model.

The loss function used in this model is a mean squared error (MSE), a common choice for regression problems, where the goal is to minimize the difference between the predicted and actual values. The model is trained using the fit() function, which takes in the training and validation data, the number of epochs, and the batch size. In this case, the model is trained for 2000 epochs with a batch size of 32, which means that the model weights are updated after every 32 samples. The verbose parameter is set to 2, meaning the training progress will be displayed on the screen. The validation data monitors the model's performance throughout the training and prevents overfitting.

This architecture is suitable for a regression problem with three input features and one output variable. The increasing and decreasing number of neurons in the hidden layers can help the model capture different levels of abstraction and avoid overfitting. The lower learning rate and larger batch size can help the optimizer converge better and avoid overfitting as well.

The main objective is to improve the neural network model to increase its generalizability and adaptability to various robotic arm configurations and tasks. Therefore, different approaches are considered, such as increasing the sophistication of neural network architectures, incorporating reinforcement learning techniques, or integrating transfer learning techniques. Furthermore, it has been suggested that the training data set be enriched to include a greater variety of scenarios and challenges, allowing the model to learn more robust and generalizable patterns.

3.5   Detection of Characters Using YOLO Neural Network

To identify characters from the Braille alphabet, the YOLO model was implemented to detect letters written in Braille in images and videos. The model was trained on a dataset of images and videos of letters written in Braille. To capture the images and videos, a modular support was implemented to hold a camera located in the upper view of the robotic arm to trace the braille alphabet. The robotic arm traced the letters in different sizes and positions on paper, and the camera captured the images and videos from different angles.

The image and video dataset was split into two sets: a training set and a validation set. The model was trained for 1000 iterations. The learning rate was gradually reduced as the model progressed through training. On the other hand, the performance was evaluated using the validation set. The model accuracy was 98%, meaning the model could correctly identify the braille alphabet in 98% of the images and videos in the validation set. The system works as follows:

1. The neural network first divides the image or video into cells.
2. Each cell is scanned for letters written in Braille.
3. If the neural network finds a braille letter, it predicts what letter it is.
4. The neural network predicts the letters for each cell in the image or video.

The convolutional neural network uses a set of features to predict letters. Features include the letters' size, shape, and location in the cell. This model can be used to identify the braille alphabet in various applications. For example, it can be used to help blind people read printed text. It can also be used to translate braille text to standard text. Figure 5 illustrates the detection process at the end of tracing a word.

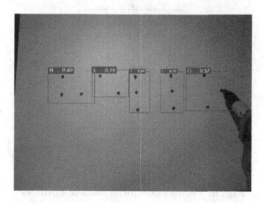

Fig. 5. Validation of the neural network for the recognition process

3.6 Logical Architecture of the System

The logical architecture of the system is illustrated in Fig. 6; this architecture describes the steps involved in generating and tracking a character, as well as the corresponding Braille and audio output. First, the system initializes and prompts the user to enter a letter or word through the graphical user interface. Then, a verification process is generated to determine if the entered character is valid (if valid, a path corresponding to the character is generated, and an audio description of the character is played). Subsequently, the robotic arm starts to trace all Brailler characters, and an image is captured using a webcam to estimate all the characters traced by the robot. Once the process is complete, the robotic arm returns to its home position.

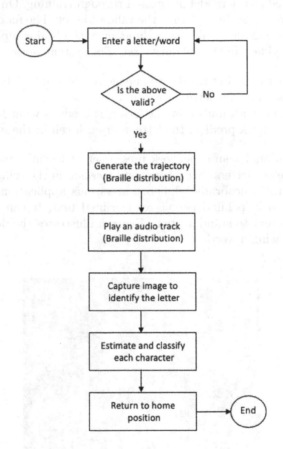

Fig. 6. Pseudocode for system operations.

4   Experimental Results

During the development of the system, components were selected considering challenges and limitations to its implementation across diverse educational contexts, including resource constraints, the availability of technical support, and adaptability. Starting from these considerations, the authors opted for 3D-printed PLA components as they provide cost-effective and visually appealing finishes. MG996R and SG90 servomotors were integrated for joint movement, and an Arduino MEGA controller facilitated seamless communication with MATLAB. Figure 7 illustrates the system's components.

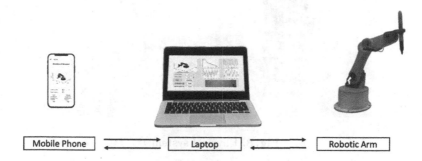

Fig. 7. Hardware resources.

The proposed robotic arm incorporates a design miming a human arm, increasing user acceptance. Compared to similar systems, it demonstrated higher efficiency and accuracy in task performance by incorporating mathematical models to analyze its kinematics and dynamics, which were validated through simulations in MATLAB. Experimental tests were conducted with ten students under the supervision of a special education specialist to evaluate the robotic arm's effectiveness as an educational tool for teaching the Braille alphabet and robotics.

To ensure user safety when interacting with the robotic arm, the initial version is presented as an unsupervised support assistant during the execution of the routines. Likewise, it is essential to specify the limit of intermediate points when addressing paths using cubic spline interpolators to optimize computational costs. Consequently, the strategy proposed in this chapter will improve the Braille system's accuracy, speed, and response time. Figure 8 shows a portable robotic arm configured to perform Braille tracings on white cardboard.

The provided model captured non-linear and complex relationships between end-of-arm and joint positions and orientations, which is crucial for optimal accuracy. In addition, this specific model features a deep neural network architecture capable of effectively learning and representing complex features. The depth and complexity of the model allow it to capture the subtleties and non-linear relationships present in the input and output data of the inverse kinematics problem. Factors such as computational efficiency and model scalability were considered, as the model must be able to execute in real-time.

Fig. 8. Physical system.

5   Graphical User Interface

Regarding the interaction of the system with the user, a graphic interface has been developed in MATLAB that integrates functions related to the choice of word or letter to draw in the Braille alphabet, graphic sections that describe the behavior of each of the joints during the execution of trajectories, simulations, and descriptive audio depending on the character to be traced (See Fig. 9).

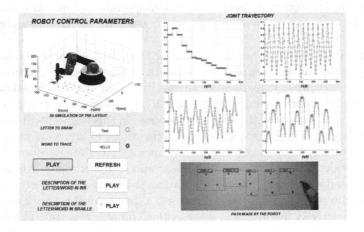

Fig. 9. Graphical user interface for computer systems.

In addition, a mobile application has been developed for Android devices using the open-source Flutter UI software development kit. This will allow intuitive operation and easy access to users who do not have a computer device (See Fig. 10).

Fig. 10. Graphical user interface for mobile devices.

6 Conclusions and Future Work

The presented research has successfully developed a robotic arm system integrated with advanced technologies, including dynamic analysis, cubic spline interpolation, artificial intelligence, and the YOLO neural network, to serve as an assistive technology tool for individuals with visual impairments. The system demonstrated high accuracy and efficiency in Braille character recognition and tracing tasks, showcasing its potential for educational and practical applications. Moreover, the logical architecture of the system was designed to ensure user-friendliness and safety, and a graphical user interface was implemented for ease of interaction. The expansion of the dataset and scenarios for YOLO neural network training can improve the system's recognition capabilities in diverse real-world settings. Finally, considering the portability and affordability of the system, its potential deployment in educational institutions and rehabilitation centers should be explored to facilitate Braille literacy and robotics education for visually impaired individuals.

References

1. Mohammed, P.S., 'Nell' Watson, E.: Towards inclusive education in the age of artificial intelligence: perspectives, challenges, and opportunities. In: Knox, J., Wang, Y., Gallagher, M. (eds.) Artificial Intelligence and Inclusive Education. PRRE, pp. 17–37. Springer, Singapore (2019). https://doi.org/10.1007/978-981-13-8161-4_2
2. Daniela, L., Lytras, M.: Educational robotics for inclusive education. Technol. Knowl. Learn. **24**(2), 219–225 (2019)
3. Aniskin, V., Korostelev, A., Lvovna, B., Kurochkin, A., Sobakina, T.: Teaching potential of integrated learning technologies Smart, Stem and Steam. Revista de la Universidad del Zulia **11**(29), 328–336 (2020)
4. Afonso, R., Soares, F., de Moura-Oliveira, P.: Innovative teaching/learning methodologies in control, automation and robotics: a short review. In: 4th International Conference of the Portuguese Society for Engineering Education (CISPEE), pp. 1–6 (2021)

5. Conde, M., Rodríguez-Sedano, F., Fernández-Llamas, C., et al.: Fostering STEAM through challenge-based learning, robotics, and physical devices: a systematic mapping literature review. Comput. Appl. Eng. Educ. **29**(1), 46–65 (2021)
6. Papadakis, S.: Robots and robotics kits for early childhood and first school age. Int. J. Interact. Mob. Technol. **14**(18), 34 (2020)
7. Bellas, F., et al.: STEAM approach to autonomous robotics curriculum for high school using the Robobo robot. In: Merdan, M., Lepuschitz, W., Koppensteiner, G., Balogh, R., Obdržálek, D. (eds.) RiE 2019. AISC, vol. 1023, pp. 77–89. Springer, Cham (2020). https://doi.org/10.1007/978-3-030-26945-6_8
8. Garg, S., Sharma, S.: Impact of artificial intelligence in special need education to promote inclusive pedagogy. Int. J. Inf. Educ. Technol. **10**(7), 523–527 (2020)
9. Oh, S.: Emergence of a new sector via a business ecosystem: a case study of Universal Robots and the collaborative robotics sector. Technol. Anal. Strategic Manage. **35**, 1–14 (2021)
10. van den Heuvel, R., Lexis, M., de Witte, L.: ZORA robot based interventions to achieve therapeutic and educational goals in children with severe physical disabilities. Int. J. Soc. Robot. **12**(2), 493–504 (2020)
11. Syriopoulou-Delli, C., Gkiolnta, E.: Robotics and inclusion of students with disabilities in special education. Res. Soc. Dev. **10**(9), e36210918238 (2021)
12. Varela-Aldás, J., Guamán, J., Paredes, B., Chicaiza, F.A.: Robotic cane for the visually impaired. In: Antona, M., Stephanidis, C. (eds.) HCII 2020. LNCS, vol. 12188, pp. 506–517. Springer, Cham (2020). https://doi.org/10.1007/978-3-030-49282-3_36
13. Gonçalves, D., Santos, G., Campos, M., Amory, A., Manssour, I.: Braille character detection using deep neural networks for an educational robot for visually impaired people. In: Anais do XVI Workshop de Visão Computacional, Evento Online, pp. 123–128 (2020)
14. Neto, I., Nicolau, H., Paiva, A.: Community based robot design for classrooms with mixed visual abilities children. In: Proceedings of the Conference on Human Factors in Computing Systems, pp. 1–12 (2021)
15. Willis, J.: Building connections with LEGO Braille bricks. Independent Education (2021)
16. Gómez, Ó.: Robotics and LOMLOE: systematic revision of robotics as inclusive tool. Hum. Rev. **13**(1), 1–13 (2022)
17. Skaraki, E.: Designing educational material to teach Braille to adult educators through the method of distance learning. Adv. Mobile Learn. Educ. Res. **3**(1), 602–609 (2023)
18. Okamoto, T., Shimono, T., Tsuboi, Y., Izumi, M., Takano, Y.: Braille block recognition using convolutional neural network and guide for visually impaired people. In: 2020 IEEE 29th International Symposium on Industrial Electronics (ISIE), pp. 487–492 (2020)
19. Alufaisan, S., Albur, W., Alsedrah, S., Latif, G.: Arabic Braille numeral recognition using convolutional neural networks. In: Bindhu, V., Tavares, J.M.R.S., Boulogeorgos, A.-A.A., Vuppalapati, C. (eds.) International Conference on Communication, Computing and Electronics Systems. LNEE, vol. 733, pp. 87–101. Springer, Singapore (2021). https://doi.org/10.1007/978-981-33-4909-4_7
20. Gálvez, M.: Sistema portable para localización de objetos y comunicación mediante audio 3D para personas invidentes (2020)
21. Abelardo, J., Montes, D.: Evaluación de las plataformas de comunicación en el aprendizaje remoto de matemáticas de los estudiantes de primero de bachillerato

de la Unidad Educativa 'Lic. Carlos Vélez Verduga' - Ecuador 2021, Universidad César Vallejo (2021)

22. Cambo, M., Ilbay, M., Sánchez, E., Zambrano, A.: Software interactivo para el apoyo del proceso y aprendizaje de las matemáticas para primero de bachillerato. Ecuadorian Sci. J. **6**(1), 32–41 (2022)

23. Máñez, C., Cervera, J.: Aplicación móvil para niños con dificultades de aprendizaje en la automatización del proceso de reconocimiento de palabras. Centro de Información Tecnológica **32**(5), 67–74 (2021)

24. Alejo, B., Aparicio, A.: La planificación de estrategias de enseñanza en un entorno virtual de aprendizaje. Revista Científica UISRAEL **8**(1), 59–76 (2021)

25. Porras, R.: Atención psicopedagógica para mejorar el aprendizaje neurosensorial en niños de 0 a 3 años, p. 2023. Universidad Estatal Península de Santa Elena, La Libertad (2023)

26. Macea, A., Laguna, D., Ruiz, N., Vega, J.: Análisis de procesos de enseñanza aprendizaje de la lectoescritura de estudiantes de grado quinto del Colegio José Celestino Mutis de Prado, Tolima. Handle.net (2022). [En línea]. Disponible en: https://hdl.handle.net/11254/1214

27. Carron, A., Arcari, E., Wermelinger, M., Hewing, L., Hutter, M., Zeilinger, M.: Data-driven model predictive control for trajectory tracking with a robotic arm. IEEE Robot. Autom. Lett. **4**(4), 3758–3765 (2019)

28. Hsiao, J., Shivam, K., Chou, C., Kam, T.: Shape design optimization of a robot arm using a surrogate-based evolutionary approach. Appl. Sci. **10**(7), 2223 (2020)

29. Siemasz, R., Tomczuk, K., Malecha, Z.: 3D printed robotic arm with elements of artificial intelligence. Procedia Comput. Sci. **176**, 3741–3750 (2020)

30. Kunze, L., Hawes, N., Duckett, T., Hanheide, M., Krajník, T.: Artificial intelligence for long-term robot autonomy: a survey. IEEE Robot. Autom. Lett. **3**(4), 4023–4030 (2018)

31. Szczepanski, R., Erwinski, K., Tejer, M., Bereit, A., Tarczewski, T.: Optimal scheduling for palletizing task using robotic arm and artificial bee colony algorithm. Eng. Appl. Artif. Intell. **113**, 104976 (2022)

32. Jin, L., Li, S., Yu, J., He, J.: Robot manipulator control using neural networks: a survey. Neurocomputing **285**, 23–34 (2018)

Spatial Shrinkage Prior: A Probabilistic Approach to Model for Categorical Variables with Many Levels

Danna Cruz-Reyes[1,2]([envelope]) [iD]

[1] Universidad del Rosario, Bogotá, Colombia
danna.cruz@urosario.edu.co
[2] Grupo de Investigación Clínica, Escuela de Medicina y Ciencias de la Salud,
Bogotá, Colombia

Abstract. One of the most commonly used methods to prevent over-fitting and select relevant variables in regression models with many predictors is the penalized regression technique. Under such approaches, variable selection is performed in a non-probabilistic way, using some optimization criterion.

A Bayesian approach to penalized regression has been proposed by assuming a prior distribution for the regression coefficients that plays a similar role as the penalty term in classical statistics: to shrink non-significant coefficients toward zero and assign a significant probability mass to non-negligible coefficients.

These prior distributions, called *shrinkage priors*, usually assume independence among the covariates, which may not be an appropriate assumption in many cases. We propose two *shrinkage priors* to model the uncertainty about coefficients that are spatially correlated.

The proposed priors are considered as an alternative approach to model the uncertainty about the coefficients of categorical variables with many levels. To illustrate their use, we consider the linear regression model. We evaluate the proposed method through several simulation studies.

Keywords: Graph of edges · Spatial correlation · Student-t distribution

1 Introduction

Regularization is a statistical methodology widely used to avoid overfitting when the model includes a large number of predictors. This method allows for the selection of relevant variables from a large set without losing computational efficiency [2,16]. The regularization technique adds a penalty term to the sum of the squared residuals of the model, grouping the coefficients that are close to zero and penalizing high-valued regression coefficients. Estimates for the regression coefficients are obtained by solving the equation

$$\min \left\{ \frac{1}{2n} |\boldsymbol{Y} - \beta_0 \boldsymbol{1} - \boldsymbol{X}\boldsymbol{\beta}|^2 + \lambda |\boldsymbol{\beta}|_q \right\} \tag{1}$$

where $|\boldsymbol{\beta}|_q = \left(\sum_{j=1}^{D} |\beta_j|^q \right)^{\frac{1}{q}}$, $\boldsymbol{Y} = (y_1, \ldots, y_n)^t$ is an n-dimensional response vector, $\boldsymbol{1}$ denotes the $n \times 1$ vector of ones, $\boldsymbol{X}$ is an $n \times D$ matrix of covariates, $\beta_0 \in \mathbb{R}$ is the intercept, $\boldsymbol{\beta} = (\beta_1, \ldots, \beta_D)^t$ is the D-dimensional vector of regression coefficients, and λ is the penalty parameter.

The value of λ is proportional to the level of shrinkage to zero of the coefficients. If $\lambda = 0$, we recover the traditional least squares estimates. There are several types of penalties defined by the value of q. If $q = 1$, we obtain the least absolute shrinkage and selection operator (Lasso) [16]. In this case, we add an L_1 penalty to the model equal to the absolute value of the coefficients' magnitude. This regularization criterion may provide sparse models as some coefficients can be shrunk toward zero and eliminated from the model. If $q = 2$, the Ridge's penalty proposed by [6] is obtained. In this case, all coefficients are shrunk by the same factor but none are eliminated from the model.

In Bayesian statistics, regularization and variable selection problems have been addressed as a prior specification problem. Various authors have discussed variable selection procedures, which can be found in references such as [3,8, 10,15,18]. One common approach for variable selection is to consider a finite mixture of distributions as the prior, as done in the spike-slab method [3].

Bayesian methods for regularization involve constructing a prior distribution for the regression coefficients that plays a role similar to the penalty term in Eq. (1). This prior shrinks coefficients that are not significant towards zero and assigns a significant probability mass to non-negligible coefficients. Bayesian methods offer several advantages, such as simultaneously estimating the penalty parameter with the model, which adds flexibility to the selection process [17]. Additionally, parameter estimation can be done using MCMC methods, which are generally robust to non-convex or multimodal penalties, conditions that often pose challenges for classical methods.

Bayesian shrinkage priors corresponding to the Lasso, grouped Lasso, Elastic nets, and Fused Lasso have been proposed by [5,9,11,13]. In these shrinkage prior formulations, the penalties correspond to specific choices of priors and are expressed as a scale-mixture of normal distributions. For example, in the linear regression model, the Ridge's penalty is equivalent to assuming a Gaussian prior distribution for $\boldsymbol{\beta}$ centered around zero, with the standard deviation defined as a function of λ. The Lasso penalty is recovered when the prior distribution for $\boldsymbol{\beta}$ is the Laplace distribution with a mean of zero, and the scale parameter depends on λ [7]. For large values of λ, only the most influential covariates are maintained in the model. These distributions are referred to as "shrinkage priors".

However, all of these methods assume independence among the variables or a known dependence structure. These assumptions can be strong in many contexts, and if not properly accounted for, they can lead to poor predictions.

The dependence structure between variables plays an important role in predicting the response, especially when dealing with a large number of predictors.

It is desirable to select the variables that significantly affect the response. To address correlated covariates in a normal linear model, [12] propose a Bayesian regularization approach that characterizes the dependence structure among the covariates through a graph Laplacian matrix.

Better predictive capacity of the model can be obtained when the dependence structure is modeled, as it enables borrowing information across variables and overcomes collinearity [12]. This is a problem that frequently occurs when the model includes categorical variables with many levels [1,4]. Such variables are represented using dummy coding, which requires the introduction of indicator variables representing each level. If the number of levels is high, the learned coefficients become very unstable, making it difficult to interpret the results. Aggregating these levels into higher-level categories is desirable, but it is not clear how to do so in a way that results in an interpretable and statistically efficient model.

One response to this problem is provided by [4], who proposed a Lasso-constrained regression approach for analysis of variance to collapse levels of a categorical covariate. [14] proposed a Bayesian regularization method to aggregate the levels of covariates based on the effect fusion prior. This prior is a modification of the spike-slab prior that considers all level effects as well as their differences, allowing for sparsity and clustering of similar coefficients. Another approach for clustering the levels of a categorical variable is considered by [1], who introduced a random partition model for clustering the coefficients.

Although some existing shrinkage priors take into account the correlation among the covariates, none of them consider the correlation induced by the spatial neighbor structure that may occur, for example, in credit risk analysis or when predicting house rent prices where geographical location is an important feature. Motivated by this, our goal is to propose a shrinkage prior distribution to model categorical predictors whose levels are spatially correlated. In addition to sparsity, this prior also has the fusion property, clustering the covariate levels that share a similar effect.

We represent these levels as spatial random effects $\boldsymbol{\theta} = (\theta_1, \ldots, \theta_r)$, where the neighboring structures will be represented by a graph. Here, r is the number of different categories and θ_i is the effect of the ith level of the categorical covariate. The nodes of the graph represent the random level effects $\boldsymbol{\theta}$, and the edges connect neighboring level effects. Similar to [12], we assume a multivariate normal prior for $\boldsymbol{\theta}$, with the covariance matrix depending on the random weights associated with the edges connecting neighbor nodes. The novelty lies in the construction of the prior for these random edge weights.

Our method explicitly characterizes the dependency structure between two level effects in $\boldsymbol{\theta}$ through the weights of edges that connect such levels. The variability of each level effect is defined by the weights of the edges that are incident to that particular node.

Let $\mathcal{G} = (\mathcal{V}, \mathcal{E})$ be an undirected graph, where $\mathcal{V} = v_1, \ldots, v_n$ is the set of vertices or nodes representing the levels of our categorical variable, and $\mathcal{E}$ is the set of p edges connecting unordered pairs of distinct vertices, representing the adjacency relationship among levels. The edge connecting $v_i \in \mathcal{V}$ and $v_j \in \mathcal{V}$ is denoted by $[ij]$. We assume that the edges are undirected, implying that $[ij] = [ji]$. If two nodes v_i and $v_j \in \mathcal{V}$ are connected, this will be denoted by $v_i \sim v_j$. When $v_i \in \mathcal{V}$ is a node in the edge $[ij] \in \mathcal{E}$, we say that the edge is incident on v_i.

Associated with the original graph $\mathcal{G}$, we define the "graph of edges" $\mathcal{L}(\mathcal{G})$. The graph of edges represents the adjacency relationship among the edges of the original graph $\mathcal{G}$. The nodes in $\mathcal{L}(\mathcal{G})$ are the edges $[ij] \in \mathcal{E}$ connecting the nodes v_i and v_j, with $i \neq j$. The edges in $\mathcal{L}(\mathcal{G})$ are also determined by the topology of $\mathcal{G}$. Two nodes $[ij]$ and $[kl]$ in $\mathcal{L}(\mathcal{G})$ are adjacent if and only if the edges $[ij]$ and $[kl]$ are incident on a common vertex. This means that the pair of neighboring edges must be of the form $[ij]$ and $[jk]$ for some $v_j \in \mathcal{V}$.

Let $\mathcal{I}_i = [ik] \in \mathcal{E}, v_k \in \mathcal{V}$ be the set of edges incident on area i.

2 Proposed Model

Consider a sample of n subjects independently selected from the population. Let $\boldsymbol{Y} = (y_1, \ldots, y_n)^t$ denote the vector of dependent continuous variables, where $y_i \in \mathbb{R}$ is the response for the ith subject. We assume that D covariates are measured, and let $\boldsymbol{X}i = (xi, 1, \ldots, x_{i,D})$ be the vector of covariates for the ith subject, generating a design matrix $\boldsymbol{X} = (\boldsymbol{X}1^t, \ldots, \boldsymbol{X}n^t)^t$ of size $n \times D$. These covariates may include both quantitative and dummy encoding of categorical variables with a few levels. Let $\boldsymbol{\beta} = (\beta_1, \ldots, \beta_D)^t$ denote the D-dimensional vector of regression coefficients that are assumed to be the same for all subjects.

Now, consider a categorical variable $\boldsymbol{Z}$ with a large number r of levels. Following [1], we represent this variable using an undirected graph $\mathcal{G} = (\mathcal{V}, \mathcal{E})$, where each level of $\boldsymbol{Z}$ is associated with a vertex in $\mathcal{V}$ and the edges in set $\mathcal{E}$ connect pairs of neighboring vertices or levels in $\mathcal{V}$. The neighborhood structure in $\mathcal{G}$ depends on the categorical variable's characteristics. The graph $\mathcal{G}$ is complete if $\boldsymbol{Z}$ is a nominal variable. If $\boldsymbol{Z}$ is an ordinal variable, the structure of $\mathcal{G}$ is simplified as we only connect levels following their natural ordering. In the case where our categorical feature is related to some spatial measurement, the graph $\mathcal{G}$ may represent a map, where vertices denote regions on the map, and edges connect regions that are spatial neighbors.

Let $\boldsymbol{Z}i = (zi1, \ldots, z_{ir})$ denote the one-hot encoding of $\boldsymbol{Z}$, indicating the vertex to which subject i belongs. Thus, for subject i, the coordinate $z_{ir} = 1$ if it belongs to vertex r, and it is zero otherwise. Let $\tilde{\boldsymbol{Z}}$ be the $n \times r$ matrix $(\boldsymbol{Z}1^t, \ldots, \boldsymbol{Z}n^t)^t$. Consider the r-dimensional vector $\boldsymbol{\theta} = (\theta1, \ldots, \theta r)^t$, where θ_r is the regression coefficient associated with the rth level of $\boldsymbol{Z}$ and is shared by all responses Y_i in vertex r.

We assume that the vertex effect is additive, and for all subject i, there is a linear relationship between y_i, $\boldsymbol{X}_i$, and $\boldsymbol{Z}_i$ given by:

$$y_i = \boldsymbol{X}_i\boldsymbol{\beta} + \boldsymbol{Z}_i\boldsymbol{\theta} + \epsilon_i, \tag{2}$$

for $i = 1, \ldots, n$, where the errors ϵ_i are independent and identically distributed (iid) as $\epsilon_i \overset{iid}{\sim} N(0, \sigma^2)$. The matrix representation of (2) is given by:

$$Y = X\beta + \tilde{Z}\theta + e, \tag{3}$$

where $e \sim N(0, \sigma_y^2 I_n)$ and I_n denotes the identity matrix of order n. Consequently, the joint distribution for Y, given $\Psi = (X, \tilde{Z}, \beta, \theta, \sigma^2)$, is the following n-variate normal distribution:

$$Y \mid \Psi \sim N_n(X\beta + \tilde{Z}\theta, \sigma^2 I). \tag{4}$$

To complete the model specification for the regression coefficients, we assume a prior distribution $\beta|\sigma \sim N_D(\mu_\beta, \sigma\Sigma_\beta)$, where $\mu_\beta \in \mathbb{R}^D$ is the known vector of prior means, and Σ_β is a $D \times D$ symmetric, positive definite matrix. The prior distribution of the model variance σ^2 is the inverse-gamma distribution with parameters $a > 0$ and $b > 0$, denoted by $\sigma^2 \sim IG(a, b)$.

Regarding the categorical variable Z, our goal is to reduce dimensionality by shrinking the effects of non-significant levels toward zero and clustering the effects of significant levels that share similar or equal effects. Additionally, we want to respect the neighbor structure of the levels in Z. To achieve this, we introduce a shrinkage prior for the effects θ that enables the grouping of levels that are strongly correlated. If our categorical covariate represents geographical regions, we aim to detect groups of areas where Z is expected to have the same value. If no group is detected, it suggests that the categorical variable takes different values across the entire region. The details of this prior will be discussed in the next section.

2.1 Spatial Fusion-Shrinkage Prior for θ

The spatial fusion-shrinkage prior distribution (SFS-prior) for the vector θ is inspired by the distribution defined in [12]. Let $\rho_{[ji]}$ denote the random weight for the edge $[ij]$ connecting levels i and j of the categorical covariate Z, and let ρ be the vector of all non-zero weights related to the graph $\mathcal{G}$. The prior for θ is hierarchically built by first specifying the conditional distribution of θ given the random weights ρ. Taking advantage of our representation of the categorical variable, which defines the Laplacian matrix as in [12], we assume that:

$$\theta|Q, \sigma^2 \sim N_r(0, \sigma^2 Q^{-1}), \tag{5}$$

However, we consider a slightly different structure for the $r \times r$ precision matrix Q by considering:

$$Q = \begin{bmatrix} 1 + \rho_{[1]} + \sum_{j \neq 1} |\rho_{[1j]}| & \cdots & -\rho_{[1r]} \\ -\rho_{[21]} & \cdots & -\rho_{[2r]} \\ \vdots & \ddots & \vdots \\ -\rho_{[r1]} & \cdots & 1 + \rho_{[r]} + \sum_{j \neq ir} |\rho_{[r]}| \end{bmatrix} \tag{6}$$

where $\rho_{[i]} > 0$ for all $i = 1, \ldots, r$, and the random weights $\rho_{[ij]} \in \mathbb{R}$ are such that $\rho_{[ij]} = \rho_{[ji]} \neq 0$ if $i \sim j$, and $\rho_{[ij]} = 0$ otherwise. The matrix Q given in (6) is positive definite because it is a real, symmetric, and strictly diagonally dominant matrix. However, it can be a sparse matrix as it depends on the neighborhood structure.

The elements of Q provide a good conditional interpretation for the proposed model. One of the advantages of the prior distribution in (5) is that the partial correlation can be directly calculated from the precision matrix and is provided by the values of the ρ vector. For two elements $(\theta_i, \theta_j) \in \theta$, the correlation is given by:

$$Corr(\theta_i, \theta_j | \theta_{-ij}, \rho) =$$

$$\begin{cases} -\dfrac{\rho_{[ij]}}{\sqrt{(1+\rho_{[i]}+\sum_{l \neq i} |\rho_{[il]}|)(1+\rho_{[j]}+\sum_{l \neq j} |\rho_{[jl]}|)}} & \text{if } i \sim j \\ 0 & \text{otherwise,} \end{cases}$$

To obtain a prior for θ that is able to shrink non-significant effects θ_i towards zero and cluster effects with similar values, the prior for ρ should be appropriately chosen to ensure that the random matrix Q is positive definite and captures the dependency structure defined by the nature of the data.

Besides, the prior distribution for ρ should also incorporate a cluster-type penalty to capture the marginal distribution of θ. In the approach proposed by Liu et al. (2014), the dependency structure between the fixed effects in the regression model is characterized by a full Laplacian matrix, and the prior distribution for the unknown elements ρ generates a marginal distribution for θ that is connected to the cluster-type penalty introduced by She and Owen (2010). She and Owen (2010) proposed a regularization method based on an L_1-type penalty that considers the magnitude of the coefficient and the pairwise differences between them.

However, the prior distribution for ρ proposed by Liu et al. (2014) models a general structure of dependence between the covariates, but does not specifically account for a neighborhood dependence structure, such as in the case where the categorical variable represents a geographic or spatial location. In order to address this issue, we modify the prior distribution to incorporate the "spatial feature" by considering the neighboring structure in the graph of edges that connect the levels of our categorical covariate.

Inspired by the work of Liu et al. (2014), we propose a new prior distribution for ρ that takes into account the spatial correlation between the effects θ_i and θ_j induced by the correlation among the incident edges on these nodes. We consider the adjacency relationship among the edges of the original graph $\mathcal{G}$, which is represented by the "graph of edges" $\mathcal{L}(\mathcal{G})$.

Based on this structure, we build the following prior distribution for $\boldsymbol{\rho}$:

$$p(\boldsymbol{\rho}|\gamma) = C_\theta |\boldsymbol{Q}|^{-1/2} \prod_{i=1}^{r} \rho_{[i]}^{-3/2} \exp\left\{-\frac{\gamma^2}{2\rho_{[i]}}\right\}$$

$$\prod_{i=2}\prod_{j<i} |\rho_{[ij]}|^{-3/2}$$

$$\exp\left\{-\frac{n_{[ij]}^2}{2|\rho_{[ij]}|} - \sum_{[kl]\sim[ij]} |\rho_{[kl]}|\right\}, \tag{7}$$

The prior distribution defined in (7) is also proper, and by combining it with the distribution in (5), we obtain the prior distribution of $\boldsymbol{\theta}$ given σ^2, denoted as $p(\boldsymbol{\theta}|\sigma)$:

$$p(\boldsymbol{\theta}|\sigma) = \frac{1}{\sigma^{r/2}} \exp\left[-\frac{1}{2\sigma^2}\left(\sum_{i=1}^{r} \theta_i^2 + \gamma\sigma\sum_{i=1}^{r} |\theta_i| + \sigma\sum_{i=2}^{r}\sum_{j<i}\sum_{j\sim i} n_{[ij]}|\theta_i - \theta_j|\right)\right]. \tag{8}$$

The prior distribution in (7) is proper and, if mixed with the prior in (5), we obtain a prior distribution for $\boldsymbol{\theta}$ that accommodates both sparsity and clustering. These properties are discussed in the following theorems.

THEOREM 1. *The prior distribution defined in (7) is proper.*

Proof. We have to prove that its integral over their parametric space is finite. From results in [12], we have to $|\boldsymbol{Q}| \leq 1$. Using Fubini's theorem, it follows that

$$\int |\boldsymbol{Q}|^{-1/2} \prod_{i=1}^{r} \rho_{[i]}^{-3/2} \exp\left\{-\frac{\gamma^2}{2\rho_{[i]}}\right\} \prod_{i=2}\prod_{j<i} |\rho_{[ij]}|^{-3/2}$$

$$\exp\left\{-\frac{n_{[ij]}^2}{2|\rho_{[ij]}|} - \sum_{[kl]\sim[ij]} |\rho_{[kl]}|\right\} d\boldsymbol{\rho}$$

$$\leq \int \prod_{i=1}^{r} \rho_{[i]}^{-3/2} \exp\left\{-\frac{\gamma^2}{2\rho_{[i]}}\right\} \prod_{i=2}\prod_{j<i} |\rho_{[ij]}|^{-3/2}$$

$$\exp\left\{\frac{n_{[ij]}}{2|\rho_{[ij]}|} - \sum_{[kl]\sim[ij]} |\rho_{[kl]}|\right\} d\boldsymbol{\rho}$$

$$= \prod_{i=1}^{r} \left[\int_{0}^{\infty} \rho_{[i]}^{-3/2} \exp\left\{-\frac{\gamma^2}{2\rho_{[i]}}\right\} d\rho_{[i]}\right]$$

$$\times \prod_{i=2}\prod_{j<i} \int_{-\infty}^{\infty} |\rho_{[ij]}|^{-3/2}$$

$$\exp\left\{-\frac{n_{[ij]}^2}{2|\rho_{[ij]}|} - \sum_{[kl]\sim[ij]} |\rho_{[kl]}|\right\} d\boldsymbol{\rho} \tag{9}$$

The first integral in expression (9) is finite as we are integrating the kernel of a inverse-gamma distribution. For the second integral, if $n_{ij} > 0$ it follows that

$$\prod_{i=2}^{r}\prod_{j<i}\int_{-\infty}^{\infty}|\rho_{[ij]}|^{-3/2}\exp\left\{-\frac{n_{[ij]}^2}{2|\rho_{[ij]}|}-\sum_{[kl]\sim[ij]}|\rho_{[kl]}|\right\}d\boldsymbol{\rho}$$

$$=\prod_{i=2}^{r}\prod_{j<i}\int_{-\infty}^{\infty}|\rho_{[ij]}|^{-3/2}$$

$$\exp\left\{-\frac{n_{[ij]}^2}{2|\rho_{[ij]}|}\right\}d\rho_{[ij]}\prod_{[kl]\sim[ij]}\int_{-\infty}^{\infty}\exp\left\{-n_{[kl]}|\rho_{[kl]}|\right\}d\rho_{[kl]}$$

$$\propto\prod_{i=2}^{r}\prod_{j<i}\int_{-\infty}^{\infty}|\rho_{[ij]}|^{-3/2}\exp\left\{-\frac{n_{[ij]}^2}{2|\rho_{[ij]}|}\right\}d\rho_{[ij]}\prod_{[kl]\sim[ij]}\frac{1}{n_{[kl]}}$$

$$=\prod_{i=2}^{r}\prod_{j<i}\int_{0}^{\infty}\rho_{[ij]}^{-3/2}\exp\left\{-\frac{n_{[ij]}^2}{2\rho_{[ij]}}\right\}d\rho_{[ij]}+\int_{-\infty}^{0}-\rho_{[ij]}^{-3/2}$$

$$\exp\left\{-\frac{n_{[ij]}^2}{2(-\rho_{[ij]})}\right\}d\rho_{[ij]}$$

$$\propto\prod_{i=2}^{r}\prod_{j<i}\left[n_{ij}^{-1/2}+n_{ij}^{-1/2}<\infty\right],$$

which concludes the proof.

THEOREM 2. *Assume the distributions in expressions (7) and (5). Let $c_{ij} = \mathrm{sign}(\rho_{ij})$, then, the prior distribution of $\boldsymbol{\theta}$, given σ^2, is given by*

$$p(\boldsymbol{\theta}|\sigma)=\frac{1}{\sigma^{r/2}}\exp\left[-\frac{1}{2\sigma^2}\sum_{i=1}^{r}\theta_i^2+\gamma\sigma\sum_{i=1}^{r}|\theta_i|\right.$$

$$\left.+\sigma\sum_{i=2}^{r}\sum_{\substack{j<i\\j\sim i}}n_{[ij]}[|\theta_i-\theta_j|1_{\rho_{ij}>0}+|\theta_i+\theta_j|1_{\rho_{ij}<0}]\right]$$

where $n_{[ij]}$ is the number of neighbors of the edge $\rho_{[ij]}$ in the edge graph $\mathcal{L}(\mathcal{G})$, $\gamma\in\mathbb{R}_+$ is a hyperparameter and 1_A is the indicator function of event A.

Proof. Assuming the distributions in expression (5) and (7) and integrating out ρ, it follows that the distribution of $\boldsymbol{\theta}$, given σ^2 is

$$p(\boldsymbol{\theta}|\sigma^2) = \frac{1}{\sigma^{r/2}} \int_0^\infty \prod_{i=1}^r$$

$$\exp\left\{-\frac{1}{2\sigma^2}(1+\rho_{[i]})\theta_i^2\right\} \rho_{[i]}^{-3/2} \exp\left\{-\frac{\gamma^2}{2\rho_{[i]}}\right\} d\rho$$

$$\times \int_{-\infty}^\infty \prod_{i=2}^r\prod_{j<i} \exp\left\{-\frac{1}{2\sigma^2}|\rho_{[ij]}|(\theta_i+c_{ij}\theta_j)^2\right\} |\rho_{[ij]}|^{-3/2}$$

$$\exp\left\{-\frac{n_{[ij]}^2}{2|\rho_{[ij]}|} - \sum_{[kl]\sim[ij]}|\rho_{[kl]}|\right\} d\rho.$$

As

$$a^{-1}\exp\{-a|z|\} = \int_0^\infty (2\pi)^{-1/2}t^{-3/2}\exp\left\{-\frac{z^2t}{2}\right\}\exp\left\{-\frac{a^2}{2t}\right\}dt$$

it follows that first integral in (10) is

$$\int_0^\infty \exp\left\{-\frac{1}{2\sigma^2}(1+\rho_{[i]})\theta_i^2\right\}\rho_{[i]}^{-3/2}\exp\left\{-\frac{\gamma^2}{2\rho_{[i]}}\right\}d\rho_{[i]} \propto$$

$$\exp\left\{-\frac{1}{2\sigma^2}\theta_i^2 - \frac{\gamma}{\sigma}|\theta_i|\right\}.$$

The second integral in (10) is

$$\int_{-\infty}^\infty \prod_{i=2}^r\prod_{j<i} \exp\left\{-\frac{1}{2\sigma^2}|\rho_{[ij]}|(\theta_i+c_{ij}\theta_j)^2\right\} |\rho_{[ij]}|^{-3/2}$$

$$\exp\left\{-\frac{n_{[ij]}^2}{2|\rho_{[ij]}|} - \sum_{[kl]\sim[ij]}|\rho_{[kl]}|\right\} d\boldsymbol{\rho}$$

$$\propto \prod_{i=2}^r\prod_{j<i} \int_{-\infty}^\infty \exp\left\{-\frac{1}{2\sigma^2}|\rho_{[ij]}|(\theta_i+c_{ij}\theta_j)^2\right\}$$

$$|\rho_{[ij]}|^{-3/2}\exp\left\{-\frac{n_{[ij]}^2}{2|\rho_{[ij]}|}\right\}d\rho_{[ij]} \times \prod_{[kl]\sim[ij]}\frac{1}{n_{[kl]}}$$

$$\propto \prod_{i=2}^r\prod_{j<i} \exp\left\{-n_{[ij]}|\theta_i+c_{ij}\theta_j|\frac{1}{\sigma}\right\}$$

$$= \prod_{i=2}^r\prod_{j<i} \exp\left\{-\frac{n_{[ij]}}{\sigma}[|\theta_i-\theta_j|1_{\rho_{ij}>0} + |\theta_i+\theta_j|1_{\rho_{ij}<0}]\right\},$$

which concludes the proof.

This prior distribution, denoted as pSFS-Prior (positive spatial fusion-shrinkage prior), accommodates both sparsity and clustering properties. It captures the interplay between the magnitude of the effects θ_i, the sparsity-inducing penalty term $\gamma\sigma$, and the pairwise differences between the effects. The constant C_θ ensures that the distribution is properly normalized.

In the general case, the prior distribution is denoted as SFS-Prior (spatial fusion-shrinkage prior). It provides a way to model the dependency structure among the effects θ while promoting sparsity and clustering.

The Geometry of the Proposed Priors for θ. The geometric properties of the contour curves associated with the prior distribution of θ provide insights into its ability to cluster similar values and shrink non-significant ones towards zero. Contour curves with a diamond shape, whose vertices lie on the horizontal and vertical axes, promote sparsity as they correspond to coordinates equal to zero. An example of such a diamond contour curve is the Lasso-prior (Fig. 1(c)). If the vertices of this diamond contour curve lie on the line $\theta_1 = \pm\theta_2$, as shown in Fig. 1(d), the prior favors the clustering of effects. The contour curve of the GL-prior proposed by [12] has an octagonal shape (Fig. 1(e)), indicating that it exhibits both sparsity and grouping properties.

The prior distributions for θ given in (8) and (8) incorporate both sparsity, with the term $|\theta_i|$, and grouping, with the terms $|\theta_i - \theta_j|$ and $|\theta_i + \theta_j|$. The constants γ and $n_{[ij]}$ reflect the degree of sparsity and grouping induced by these priors, respectively. Panels (a) and (b) of Fig. 1 show the contour plots for the SFS-prior and pSFS-prior, respectively, providing geometric evidence of these characteristics in the two-dimensional case.

We assume $\sigma^2 = 1$ and vary γ and the number $n_{[ij]}$ of neighbors of the edge $\rho_{[ij]}$ in the edge graph $\mathcal{L}(\mathcal{G})$. The black lines in Panels (a) and (b) of Fig. 1 represent the contour curves of the SFS-prior and pSFS-prior, respectively, for $\gamma = n_{[ij]} = 1$. With this parametrization, both priors promote sparsity and clustering. The dashed (blue) lines (dot-dashed (red) lines, respectively) represent the contour curves of the SFS-prior and pSFS-prior for $\gamma = 1$ ($\gamma = 2, 4, 6$, respectively) and $n_{[ij]} = 2, 4, 6$ ($n_{[ij]} = 1$, respectively). From the dashed (blue) lines, it can be observed that, for a fixed $\gamma = 1$, as $n_{[ij]}$ increases, the contour curves of the SFS-prior assume the shape of a grouping prior distribution, promoting the clustering of neighboring effects. Similarly, for a fixed $n_{[ij]}$, as γ increases, the dot-dashed (red) lines exhibit the shape of the Lasso-prior, favoring sparsity.

3   Posterior Inference

Assuming the specified prior distributions, we can derive the full conditional distribution of each parameter. Given the observed design matrix $D = \{X, \tilde{Z}\}$ and the sample of response variables $Y = (Y_1, \ldots, Y_n)$, the full conditional distributions are as follows:

β:

$$\beta | D, Y, \theta, \rho, \sigma^2 \sim N_D(\mu_\beta^*, \Sigma_\beta^*),$$

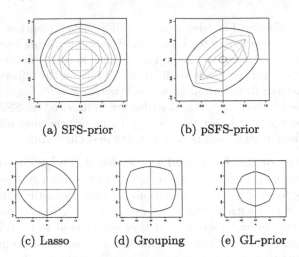

(a) SFS-prior   (b) pSFS-prior

(c) Lasso   (d) Grouping   (e) GL-prior

Fig. 1. Bidimensional contour plot of $-\log(p(\theta|\sigma^2))$ for the proposed priors, SFS-prior (a) and pSFS-prior (b), Lasso prior (c), Grouping prior (d) and GL-prior (e).

where

$$\Sigma_\beta^* = \left(X^TX + \frac{1}{\sigma^2}\Sigma_\beta^{-1}\right)^{-1},$$

$$\mu_\beta^* = \Sigma_\beta^* \left(X^TY + \frac{1}{\sigma^2}\Sigma_\beta^{-1}\mu_\beta\right).$$

σ^2:

$$\sigma^2|D,Y,\theta,\rho,\beta \sim IG\left(a+\frac{n}{2},b^*\right).$$

$$b^* = b + \frac{1}{2}\left(Y - X\beta - \tilde{Z}\theta\right)^T\left(Y - X\beta - \tilde{Z}\theta\right)$$

θ:

$$\theta|D,Y,\beta,\rho,\sigma^2 \sim N_r(\mu_\theta^*,\Sigma_\theta^*),$$

where

$$\Sigma_\theta^* = \left(\sigma^2Q + \frac{1}{\sigma^2}I_r\right)^{-1},$$

$$\mu_\theta^* = \Sigma_\theta^*\left(\sigma^2\tilde{Z}^T(Y - X\beta) + \frac{1}{\sigma^2}0\right).$$

ρ:

$$\rho|D,Y,\beta,\theta,\sigma^2 \sim \text{Dirichlet}(c^*),$$

where c^* is a vector of shape parameters defined based on the neighborhood structure of the graph $\mathcal{G}$ and the values of θ.

These full conditional distributions can be used within a Markov chain Monte Carlo (MCMC) algorithm to sample from the joint posterior distribution of the parameters.

- The posterior full conditional distributions (fcd) of σ^2 is the Inverse-Gamma distribution $\sigma|\boldsymbol{\beta},\boldsymbol{\theta},\boldsymbol{Q},\boldsymbol{Y},\boldsymbol{D} \sim$

$$IG\left(a + \frac{1}{2}(n + D + r), b + \frac{1}{2}(\tilde{Y} + \tilde{\beta} + \tilde{\theta})\right),$$

where $\tilde{Y} = (\boldsymbol{Y} - (\tilde{\boldsymbol{Z}}\boldsymbol{\theta} + \boldsymbol{X}\boldsymbol{\beta}))^t(\boldsymbol{Y} - (\tilde{\boldsymbol{Z}}\boldsymbol{\theta} + \boldsymbol{X}\boldsymbol{\beta}))$, $\tilde{\beta} = (\boldsymbol{\beta} - \mu_\beta)^t\Sigma_\beta(\boldsymbol{\beta} - \mu_\beta)$ and $\tilde{\theta} = \boldsymbol{\theta}^t\boldsymbol{Q}^{-1}\boldsymbol{\theta}$.

- The posterior fcd of $\boldsymbol{\beta}$ is the normal distribution

$$\boldsymbol{\beta}|\sigma,\boldsymbol{Q},\boldsymbol{\theta},\boldsymbol{Y},\boldsymbol{D} \sim N(\tilde{\boldsymbol{S}}(\boldsymbol{X}^t(\boldsymbol{Y} - \tilde{\boldsymbol{Z}}\boldsymbol{\theta}) + \Sigma_\beta^{-1}\mu_\beta), \sigma\tilde{\boldsymbol{S}})$$

where $\tilde{\boldsymbol{S}} = (\boldsymbol{X}^t\boldsymbol{X} + \Sigma_\beta^{-1})^{-1}$.

- The posterior fcd of $\boldsymbol{\theta}$ is the normal distribution

$$\boldsymbol{\theta}|\sigma,\boldsymbol{Q},\boldsymbol{\beta},\boldsymbol{Y},\boldsymbol{D} \sim N_r(\tilde{\boldsymbol{S}}'(\tilde{\boldsymbol{Z}}^t(\boldsymbol{Y} - \boldsymbol{X}\boldsymbol{\beta})), \sigma\tilde{\boldsymbol{S}}')$$

where $\tilde{\boldsymbol{S}}' = (\tilde{\boldsymbol{Z}}^t\tilde{\boldsymbol{Z}} + \boldsymbol{Q})^{-1}$.

Considering the SFS-prior it is simpler to separately sample from the posterior distribution of $\rho_{[i]}$ and $\rho_{[ij]}$. Let $c_{ij} = sign(\rho_{ij})$. It follows from (5) that, given $\boldsymbol{\theta}$ and σ^2, $\rho_{[i]}$ and $\rho_{[ij]}$ are independent. Thus the posterior fcd of $\rho_{[i]}$ is given by

$$p(\rho_{[i]}|\sigma,\boldsymbol{\theta},\boldsymbol{Y},\boldsymbol{D}) \propto \rho_{[i]}^{-3/2}\exp\left\{-\frac{1}{2\sigma^2}\rho_{[i]}\theta_i^2 - \frac{\gamma^2}{2\rho_{[i]}}\right\}$$

which is a inverse-Gaussian distribution with mean $\gamma\sigma|\theta_i|^{-1}$ and shape parameter γ^2 denoted by $\rho_{[i]}|\sigma,\boldsymbol{\theta},\boldsymbol{Y},\boldsymbol{D} \sim IGaussian(\gamma\sigma|\theta_i|^{-1}, \gamma^2)$. The posterior fcd of $\rho_{[ij]}$ assuming the SFS-Prior and pSFS-Prior have unknown closed-form and are given, respectively, by

$$p(\rho_{[ij]}|\sigma,\theta,\boldsymbol{Y},\boldsymbol{D})$$

$$\propto \exp\left\{-\frac{1}{2\sigma^2}|\rho_{[ij]}|(\theta_i + c_{ij}\theta_j)^2\right\}|\rho_{[ij]}|^{-3/2}\exp\left\{-\frac{n_{[ij]}^2}{2|\rho_{[ij]}|} - \sum_{[kl]\sim[ij]}|\rho_{[kl]}|\right\}$$

$$\propto |\rho_{[ij]}|^{-3/2}\exp\left\{-\frac{1}{2\sigma^2}|\rho_{[ij]}|(\theta_i + c_{ij}\theta_j)^2 - \frac{n_{[ij]}^2}{2|\rho_{[ij]}|} - \sum_{[kl]\sim[ij]}|\rho_{[kl]}|\right\}.$$

and

$$p(\rho_{[ij]}|\sigma,\theta,Y,D)$$

$$\propto \exp\left\{-\frac{1}{2\sigma^2}\rho_{[ij]}(\theta_i-\theta_j)^2\right\}\rho_{[ij]}^{-3/2}\exp\left\{-\frac{n_{[ij]}^2}{2\rho_{[ij]}}-\sum_{[kl]\sim[ij]}\rho_{[kl]}\right\}$$

$$\propto |\rho_{[ij]}|^{-3/2}\exp\left\{-\frac{1}{2\sigma^2}|\rho_{[ij]}|(\theta_i+c_{ij}\theta_j)^2-\frac{n_{[ij]}^2}{2\rho_{[ij]}}-\sum_{[kl]\sim[ij]}\rho_{[kl]}\right\}.$$

To sample from the posterior, we propose the following Gibbs sampler scheme with Metropolis-Hasting step.

Algoritmo 1: MCMC scheme to sample from the posterior distribution.

---

1 **Input:** D, Y ;
2 initialization$(\sigma^{2(0)},\beta^{(0)},\theta^{(0)},\rho^{(0)})$;
3 **for** $t=1$ to T **do**
4     $\sigma^{2(t)} \sim p(\sigma|\beta^{(t-1)},\theta^{(t-1)},Q^{(t-1)},Y,D)$;
5     $\beta^{(t)} \sim p(\beta|\sigma^{(t)},\theta^{(t-1)},Q^{(t-1)},Y,D)$;
6     $\theta^{(t)} \sim p(\theta|\sigma^{(t)},\beta^{(t)},Q^{(t-1)},Y,D)$;
7   **for** $i=1$ to n **do**
8         $\rho_{[i]} \sim p(\rho_{[i]}|,\sigma^{(t)},\theta^{(t)},Y,D)$;
9   **end**
10   **for** *[ij]*=1 to p **do**
11         $\rho'_{[ij]} \sim g(\rho_v)$;
12   **if** $\alpha(\rho'_{[ij]}|\rho_{[ij]}^{t-1}) \geq unif(0,1)$ **then**
13             $\rho_{[ij]}^t = \rho'_{[ij]}$ **else**
14             $\rho_{[ij]}^t = \rho_{[ij]}^{t-1}$
15   **end**
16         $Q^{(t)} = f(\rho_{[i]},\rho_{[ij]})$
17   **end**
18   **end**
19 **end**

---

4 Simulation Study

To evaluate the performance of the proposed approaches for clustering and shrinking non-significant effects, we conducted a simulation study. We considered a regression model with a categorical variable related to a spatial effect. The study was conducted under three different scenarios. In Scenario 1, there were no spatial clusters, and the effects were independently generated from a uniform distribution. In Scenarios 2 and 3, three clusters were considered, and the effects were fixed accordingly. We compared the SFS-prior and pSFS-prior with other models, including the unpenalized maximum likelihood (MaxLik), the PPRM, Lasso, grouping, EffectFusion, and GL-prior methods.

For each scenario, we collected 500 MCMC iterations after a burn-in of 100 iterations. We compared the models using various model selection criteria, including DIC, AIC, and RMSE. We also calculated the false positive and false negative rates related to cluster estimation. By incorporating these evaluation measures, we aimed to comprehensively assess the performance of the different models, considering both their fit to the data and their ability to identify meaningful clusters. This multifaceted approach allowed us to gain a comprehensive understanding of the strengths and limitations of each modeling approach.

The results of the simulation study showed that the SFS-prior and pSFS-prior had competitive performance compared to other models. In Scenario 1, where there were no clusters, MaxLik had the best performance. However, the shrinkage priors that accounted for spatial association among the categorical effects (SFS-prior, pSFS-prior, and GL-prior) provided comparable results. The SFS-prior showed a higher false positive rate, indicating that it tended to force clustering even when it was not necessary.

In Scenarios 2 and 3, where there were clusters, the pSFS-prior showed the best model fitting according to DIC and AIC. The SFS-prior and GL-prior also had reasonable performance in these scenarios. Overall, the proposed approaches were competitive even in scenarios that did not favor their specific features.

These results indicate that the SFS-prior and pSFS-prior can effectively cluster similar effects and shrink non-significant effects towards zero in the presence of spatial association. They provide a flexible and robust framework for modeling categorical predictors with spatial correlation.

Figure 2 illustrates the clustering results obtained from the posterior estimates of θ_i in Scenario 3. To determine the clusters, we calculated the average of the posterior means for each State and compared them using highest density posterior intervals for $\theta_i - \theta_j$ with a probability of 0.95. If the intervals did not overlap, we considered the averages to be significantly different and assigned the corresponding States to different clusters. In this analysis, all models except for MaxLik, Grouping, and GL-Prior methods successfully recovered the original clustering structure. This observation is consistent with the results presented in Table 1, which indicate that EffectFusion, PPRM, GL-Prior, pSFS-Prior, and SFS-Prior exhibited similar performance in this scenario. The ability of these models to capture the underlying clustering pattern suggests their effectiveness in identifying spatial associations among the categorical effects.

Table 1. Model comparison criteria for all fitted models.

Model	MSE	DIC	AIC	FN	FP
Scenario 1					
Maxlik	0.30	**4436.02**	**4463.21**		0.00
Lasso	0.35	4728.32	4756.32		0.00
Grouping	0.48	6245.62	6245.66		0.00
EffectFussion	0.52	6529.12	5821.72		0.00
PPRM	0.51	7764.17	7803.02		0.00
GL-prior	0.35	4834.09	7901.60		0.00
pSFS-prior	**0.32**	4992.61	5649.61		0.00
SFS-prior	0.34	4464.35	4914.63		**0.20**
Scenario 2					
Maxlik	0.67	11014.22	11089.51	1.00	0.00
Lasso	0.29	10604.51	10604.52	0.75	0.00
Grouping	**0.11**	11345.32	11345.53	0.40	0.00
EffectFussion	0.22	11345.21	11346.23	0.40	0.20
PPRM	0.13	**7163.82**	12725.62	0.25	**0.15**
GL-prior	0.12	8733.91	13362.92	0.25	0.25
pSFS-prior	0.12	7279.81	**10089.52**	**0.20**	0.25
SFS-prior	0.13	8940.71	11032.22	0.25	0.25
Scenario 3					
Maxlik	0.57	17521.82	12522.93	1.00	0.00
Lasso	0.23	12057.67	12557.62	1.00	0.00
Grouping	0.23	11057.68	11157.82	1.00	0.15
EffectFussion	0.11	1284.40	3598.04	1.00	0.15
PPRM	0.11	1264.40	3358.54	0.50	0.25
GL-prior	0.10	1107.71	1799.11	0.50	0.25
pSFS-prior	0.11	1259.74	2893.99	0.25	0.25
SFS-prior	**0.10**	**935.83**	**1658.72**	**0.25**	**0.25**

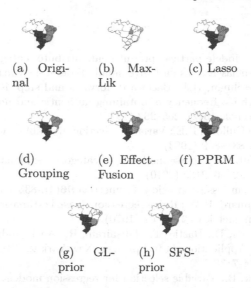

(a) Original

(b) Max-Lik

(c) Lasso

(d) Grouping

(e) Effect-Fusion

(f) PPRM

(g) GL-prior

(h) SFS-prior

Fig. 2. Clusters of the spatial effects for each model.

5   Conclusion

In conclusion, we proposed a novel approach for analyzing categorical variables with a large number of levels and correlated effects. The approach is based on a shrinkage prior inspired by the group Lasso and GL-prior penalties used for geographical predictors. We compared the performance of our shrinkage prior with other popular penalties such as Lasso, grouping, effect fusion, and the GL-prior, as well as with the non-penalized approach commonly used for categorical variables.

Our results demonstrated that the proposed shrinkage prior outperformed several existing methods, including Lasso, grouping, effect fusion, and the GL-prior. The performance of the shrinkage prior was even superior to the non-penalized approach, highlighting its effectiveness in handling high-dimensional categorical variables with correlated effects.

We presented two variants of the shrinkage prior: one that considers clusters of the same sign and another that allows for different signs. This flexibility is important as it allows for different applications in spatial statistics, where it may be desirable to consider only correlations of the same sign.

One of the key advantages of our approach is that it simultaneously estimates the values of the precision matrix corresponding to the edge weights in the underlying graph. However, further investigation is needed to optimize the behavior of the edge weights and explore their implications in more detail.

Overall, our proposed shrinkage prior offers a promising solution for analyzing categorical variables with correlated effects, providing improved model performance and valuable insights into the underlying structure of the data.

References

1. Criscuolo, T.L.: Modelo partição produto para atributos categóricos. Ph.D. thesis, Instituto de Ciências Exatas da Universidade Federal de Minas Gerais (2019)
2. Derksen, S., Keselman, H.J.: Backward, forward and stepwise automated subset selection algorithms: frequency of obtaining authentic and noise variables. Br. J. Math. Stat. Psychol. **45**(2), 265–282 (1992)
3. George, E.I., McCulloch, R.E.: Variable selection via Gibbs sampling. J. Am. Stat. Assoc. **88**(423), 881–889 (1993)
4. Gertheiss, J., Tutz, G.: Sparse modeling of categorial explanatory variables. Ann. Appl. Stat. **4**(4), 2150–2180 (2010)
5. Hans, C.: Bayesian lasso regression. Biometrika **96**(4), 835–845 (2009)
6. Hoerl, A.E., Kennard, R.W.: Ridge regression: biased estimation for nonorthogonal problems. Technometrics **12**, 55–67 (1970)
7. James, G., Witten, D., Hastie, T., Tibshirani, R.: An Introduction to Statistical Learning: With Applications in R. Springer, New York (2013). https://doi.org/10.1007/978-1-4614-7138-7
8. Kuo, L., Mallick, B.: Variable selection for regression models. Indian J. Stat. Ser. B **60**, 65–81 (1998)
9. Kyung, M., Gill, J., Ghosh, M., Casella, G.: Penalized regression, standard errors, and Bayesian lassos. Bayesian Anal. **5**(2), 369–411 (2010)
10. Li, F., Zhang, N.R.: Bayesian variable selection in structured high-dimensional covariate spaces with applications in genomics. J. Am. Stat. Assoc. **105**(491), 1202–1214 (2010)
11. Li, Q., Lin, N.: The Bayesian elastic net. Bayesian Anal. **5**(1), 151–170 (2010)
12. Liu, F., Chakraborty, S., Li, F., Liu, Y., Lozano, A.C.: Bayesian regularization via graph Laplacian. Bayesian Anal. **9**(2), 449–474 (2014)
13. Park, T., Casella, G.: BAMLSS: the Bayesian Lasso. J. Am. Stat. Assoc. **103**(482), 681–685 (2008)
14. Pauger, D., Wagner, H.: Bayesian effect fusion for categorical predictors. Bayesian Anal., 341–369 (2017)
15. Smith, M., Kohn, R.: Nonparametric regression using Bayesian variable selection. J. Econom. **75**(2), 317–343 (1996)
16. Tibshirani, R.: Regression shrinkage and selection via the lasso. J. Royal Stat. Soc. (Ser. B) **58**, 267–288 (1996)
17. Van Erp, S., Oberski, D., Mulder, J.: Shrinkage priors for Bayesian penalized regression. J. Math. Psychol. **89**, 31–50 (2019)
18. Vannucci, M., Stingo, F., Berzuini, C.: Bayesian Models for Variable Selection That Incorporate Biological Information. Oxford University Press (2012). 9780199694587. Publisher Copyright: © Oxford University Press 2011. All rights reserved. Copyright: Copyright 2018 Elsevier B.V., All rights reserved

Author Index

© The Editor(s) (if applicable) and The Author(s), under exclusive license
to Springer Nature Switzerland AG 2024
A. D. Orjuela-Cañón et al. (Eds.): ColCACI 2023, CCIS 1865, p. 171, 2024.
https://doi.org/10.1007/978-3-031-48415-5